The Digital Twin of Humans

Iris Gräßler · Günter W. Maier · Eckhard Steffen ·
Daniel Roesmann

Editors

The Digital Twin of Humans

An Interdisciplinary Concept of Digital
Working Environments in Industry 4.0

 Springer

Editors
Iris Gräßler 🆔
Heinz Nixdorf Institute
Paderborn University
Paderborn, North Rhine-Westphalia
Germany

Günter W. Maier 🆔
Department of Psychology
Bielefeld University
Bielefeld, North Rhine-Westphalia
Germany

Eckhard Steffen 🆔
Paderborn Center for Advanced Studies
Paderborn University
Paderborn, North Rhine-Westphalia
Germany

Daniel Roesmann
Heinz Nixdorf Institute
Paderborn University
Paderborn, North Rhine-Westphalia
Germany

ISBN 978-3-031-26106-0   ISBN 978-3-031-26104-6 (eBook)
https://doi.org/10.1007/978-3-031-26104-6

Preface

Industry 4.0 is an omnipresent keyword in Germany for the digitalisation of industrial work. Or should we say: ...has been an omnipresent keyword. In the meantime, we find ourselves in the digital transformation of society with far-reaching consequences in all areas of life. At the time of its founding in 2014, the *NRW Research College: Design of Flexible Working Environments—Human-Centered Use of Cyber-Physical Systems in Industry 4.0* was one of the first scientific programs to focus on the role of humans in this transformation process. In its first funding phase, powerful instruments for interdisciplinary basic research in this area were developed and established. In the second funding phase, the role of humans in the conception and modeling of hybrid systems was placed at the center of research activities, leading almost inevitably to the concept of the human digital twin. The leading research question of the importance and impact of a digital twin of humans in the design of cyber-physical-sociotechnical systems is considered from a variety of perspectives. This handbook summarizes the results of the research conducted in the NRW Research College during the second funding phase. The diverse contributions provide a comprehensive insight into the current state of research in this topic. We would like to thank all members of the *NRW Research College: Design of Flexible Working Environments—Human-Centered Use of Cyber-Physical Systems in Industry 4.0* for their high level of commitment and the many intensive discussions. Our special thanks go to the non-scientific partners who brought a transdisciplinary perspective to the project, resulting in very exciting research questions. We would like to thank the Ministry of Culture and Science of the federal state North Rhine-Westphalia and the Universities of Bielefeld and Paderborn without whose support this ambitious research program could not have been carried out.

Paderborn, Germany   Gregor Engels
November 2022   Chair of the NRW Research
                                              College: Design of Flexible Working
                                             Environments—Human-Centered Use
                                                   of Cyber-Physical Systems in
                                                                 Industry 4.0

                                                               Iris Gräßler
                                                            Günter W. Maier
                                                           Eckhard Steffen
                                                         Daniel Roesmann

Contents

The Effects of the Digital Twin of Humans

Contributors

Anja-Kristin Abendroth Faculty of Sociology, Bielefeld University, Bielefeld, Germany

Sarah Brunsmeier Faculty of Sociology, University Bielefeld, Bielefeld, Germany

Victoria Buchholz Social Cognitive Systems Group, Bielefeld University, Bielefeld, Germany

Chiara Cappello Department of Mathematics, Paderborn University, Paderborn, Germany

Martin Diewald Faculty of Sociology, University Bielefeld, Bielefeld, Germany

Roman Dumitrescu Heinz Nixdorf Institute, University of Paderborn, Paderborn, Germany;
Fraunhofer Institute for Mechatronic Systems Design, Paderborn, Germany

Elisa Gensler Faculty of Sociology, Bielefeld University, Bielefeld, Germany

Iris Gräßler Heinz Nixdorf Institute, Paderborn University, Paderborn, Germany

Christian Harteis Paderborn University, Paderborn, Germany

Talea Hellweg University Paderborn, Paderborn, Germany;
Paderborn University, Paderborn, Germany

Paul Hellwig Work and Organizational Psychology, Bielefeld University, Bielefeld, Germany;
CoR-Lab, Bielefeld University, Bielefeld, Germany

Paul Hemsen Chair of Personnel Economics, University Paderborn, Paderborn, Germany

Christian Koldewey Heinz Nixdorf Institute, University of Paderborn, Paderborn, Germany

Stefan Kopp Social Cognitive Systems Group, Bielefeld University, Bielefeld, Germany

Günter W. Maier Work and Organizational Psychology, Bielefeld University, Bielefeld, Germany;
CoR-Lab, Bielefeld University, Bielefeld, Germany

Jörn Steffen Menzefricke Heinz Nixdorf Institute, University of Paderborn, Paderborn, Germany

Hendrik Oestreich CoR-Lab, Bielefeld University, Bielefeld, Germany

Sarah Pilz CITEC – Bielefeld University, Bielefeld, Germany

Alexander Pöhler Heinz Nixdorf Institute, Paderborn University, Paderborn, Germany

Mareike Reimann Faculty of Sociology, Bielefeld University, Bielefeld, Germany

Daniel Roesmann Heinz Nixdorf Institute, Paderborn University, Paderborn, Germany

Ulrich Rückert CITEC – Bielefeld University, Bielefeld, Germany

Martin Schneider Chair of Personnel Economics, Paderborn University, Paderborn, Germany

Eckhard Steffen Paderborn Center of Advanced Studies, Paderborn University, Paderborn, Germany;
Department of Mathematics, Paderborn University, Paderborn, Germany

Britta Wrede Software Engineering for Cognitive Robots and Cognitive Systems, Faculty 3-Mathematics and Computer Science, University of Bremen, Bremen, Germany

Sebastian Wrede CoR-Lab, Bielefeld University, Bielefeld, Germany

Fundamentals

Introduction—The Digital Twin of Humans

Iris Gräßler, Eckhard Steffen, Günter W. Maier, and Daniel Roesmann

Abstract The necessary integration of human digital twins into the conception and description of digital working worlds in Industry 4.0 requires interdisciplinary research approaches, which are described in the chapters of this handbook. We introduce the topic and give a brief overview of this handbook, including short abstracts of the individual chapters.

1 Introduction

For decades, digital technologies have been penetrating all areas of daily life at a rapid pace in both private and professional areas. These technologies enable companies to aggregate huge volumes of data (Gräßler, 2003). The increasing usage of digital technologies has led to fundamental changes in the world of work (Kauffeld & Maier, 2020). For example, new forms of collaborative work are being created via digital platforms. Intelligent digital assistance systems support the execution of tasks. These systems are developed in a way that allows them to adapt to tasks and employees (Gräßler et al., 2020). The advanced and intelligent processing of data enables new opportunities, such as new business models, collaborative work, the improvement of current systems (Gräßler & Pöhler, 2020; Gräßler et al., 2022), and human-centred

I. Gräßler (✉) · D. Roesmann
Heinz Nixdorf Institute, Paderborn University, Fürstenallee 11, 33102 Paderborn, Germany
e-mail: iris.graessler@hni.uni-paderborn.de

D. Roesmann
e-mail: daniel.roesmann@hni.uni-paderborn.de

E. Steffen
Paderborn Center of Advanced Studies, Paderborn University, Fürstenallee 11, 33102 Paderborn, Germany
e-mail: es@uni-paderborn.de

G. W. Maier
Work and Organizational Psychology, Bielefeld University, Universitätsstraße 25, 33615 Bielefeld, Germany
e-mail: ao-psycholgie@uni-bielefeld.de

I. Gräßler et al. (eds.), *The Digital Twin of Humans*,
https://doi.org/10.1007/978-3-031-26104-6_1

work design (Mlekus et al., 2022). Currently, the potential of data analysis and data treatment has not yet been fully realised. Many experts predict that the concept of the digital twin has great potential in the context of digital transformation (Liu et al., 2021). Digital twins are digital representations of product instances, product-service systems, and human beings, and they are based on the use of cyber-physical systems. Such systems perceive their environment via suitable sensors. By linking the physical world with the digital world, new types of analyses of existing systems and the prediction of the future behaviour of such systems are possible. This means that measures can be taken at an early stage to improve a system. Existing applications show that the use of the digital twins of technical systems has advantages. Within the condition monitoring and assessment of production systems, for example, the digital twin is used for condition assessment. This allows weaknesses to be detected at an early stage, and it allows failures to be avoided (Macchi et al., 2018). Research in the field of digital twins mainly addresses technical systems (Liu et al., 2021). Socio-technical systems in the working environment of the future, such as production systems, will continue to require employees in addition to technical elements. Data on employees will also be collected in an anonymised and pseudonomised form in the context of digitalisation (Gräßler & Pöhler, 2019). The collected personal data of employees present the opportunity to create a digital representation of a human being that conforms to the definition of a digital twin (Gräßler & Pöhler, 2017). These digital twins of humans include selected characteristics and behaviours of human beings that are linked to models, information, and data (Gräßler et al., 2020). According to existing trend studies (DHL, 2019; Gartner, 2018), the digital twin of a human being is a technology that will have a significant impact on the economy, society, and people. Promising applications include, for example, the human-centred design of adaptive assistance systems that consider the capabilities and preferences of human workers (Oestreich et al., 2019; Gräßler et al., 2020; Buchholz & Kopp, 2020). However, it is important to consider the regulatory framework for the use of personal data and threats of misuse. The introduction of such technologies is an enormous challenge, and therefore the technical perspective must be enriched by considering aspects of justice (Hellwig et al., 2023; Ötting & Maier, 2018) and work autonomy (Gensler & Abendroth, 2021). A crucial aspect of digitising the working environment is determining how the data can be used optimally and how the potential misuse of the data can be counteracted (Engels, 2020). Therefore, an interdisciplinary concept of digital working environments is needed to enable the implementation of the digital twins of humans. The relevant disciplines are engineering, economics, computer science, mathematics, psychology, sociology, and education. Within this book, these viewpoints of digital twins are focused on, creating a holistic view.

2 Digital Twin of Humans

Since 2010, the term 'Work 4.0' (derived from Industry 4.0) has been strongly present in research. The research activities and the number of papers concerning these topics are increasing rapidly; this is particularly true for the field of digital twins. Within this section, the consideration of the digital twins of humans in a socio-technical system is proposed.

The origin of the idea of a digital twin can be traced back to the Apollo programme of the National Aeronautics and Space Administration (NASA) in the 1970s. Within this programme, two identical space vehicles were developed to mirror the conditions during a space mission (Boschert & Rosen, 2016). The term 'digital twin' was first introduced in 2004 by Grievers as part of the course 'Product Lifecycle Management' at the University of Michigan in the United States (Grievers, 2014).

The first definition of the term 'digital twin' was given by NASA in 2012. After this, this term was discussed in the aerospace field and then in different fields involving mechatronic systems (Grievers, 2014). A crucial field for the development and use of digital twins is product engineering. Within product and production system engineering, digital twins are designed. Data from previous products and production systems can be used. Therefore, we refer to the definition of the Scientific Society for Product Development (WiGeP) from Germany.

The WiGeP defines the digital twin in the following way: 'A Digital Twin is a digital representation of a product instance or an instance of a product-service system. This digital representation includes selected features, states and behaviour of the product instance or system. Similarly, different models, information and data are linked together within this digital representation during different life cycle phases. [...] The Digital Twin contains links between the digital master and the digital shadow' (Stark et al., 2020). The digital twin thus possesses data from the development of a product that are identical within a class, and these data are enriched by data concerning a specific product instance. Throughout this book, the production system is the system of interest that must be developed, operated, and maintained. With the use of these data, appropriate conclusions and predictions can be made about specific instances in a product's use phase.

The basis for this is the selection of suitable sensor technology. Every activity in a socio-cyber-physical system can be automatically recorded by suitable sensor technology; then, it can be stored and used as a basis for future decisions.

Within the research programme 'Design of Flexible Working Environments—Human-Centric Use of Cyber-Physical Systems in Industry 4.0', attempts are now being made to transfer the concept of the digital twin to humans. The introduction of the digital twins of humans can improve work processes and increase productivity, and it can improve the well-being and working conditions of employees. This concept can be used in different application areas such as workplace design, production scheduling, and assistance system design.

Like the digital twins of technical system elements, the digital twins of humans need a counterpart in the real world. The digital representation can be modelled by using various attributes of a human being. The digital twins of humans include at least one of the following attributes (Miller & Spatz, 2022; Grosse et al., 2017; Gräßler et al., 2020; Shengli, 2021; Engels, 2020).

- **Physical attributes**: e.g., anthropometric attributes, biomechanical attributes, eye movements, injuries;
- **Physiological attributes**: e.g., heart rate, muscle tension, brain electrophysiological signals, blink rates and timing;
- **Psychosocial attributes**: e.g., motivation, stress;
- **Mental attributes**: e.g., experience, skills, abilities, workload, intuitive bias;
- **Perceptual attributes**: e.g., auditory sensitivity, visual sensitivity, temperature sensitivity;
- **Emotional attributes**: e.g., anger, fear, sadness, shame;
- **Behavioural attributes**: e.g., interactions with the system.

Based on Miller and Spatz (2022), Grosse et al. (2017), Gräßler et al. (2020), Shengli (2021), and Engels (2020), we define the digital twin of a human being as an integrated model that enables descriptive analytics, diagnostic analytics, predictive analytics, and prescriptive analytics based on the data collected in socio-technical systems. The digital twin contains different elements and information for individualised descriptions and evaluations, with the aim of investigating one or more characteristics of a human or humans. For this purpose, the model combines established models (for example, models from occupational science, such as learning curves (Gräßler et al., 2021)) with real-time data from the physical twin.

3   Interdisciplinary Perspectives—Contributions in This Book

This book contains a number of contributions on the digitalisation of work from different disciplines. The focus of all these contributions is the use of a digital representation of the employee, the so-called digital twin of a human being. In addition to the basics, the contributions are divided into the areas of planning digital and networked work, implementing the digital twins of humans, and the effects of the digital twins of humans (see Fig. 1).

In the context of the **fundamentals**, the basis for the following chapters is created. This part illustrates changes in the digital world of work. It discusses the opportunities and risks that the digital twins of humans have for the future world of work.

In the chapter *Who Will Own Our Global Digital Twin: The Power of Genetic and Biographic Information to Shape Our Lives* by Pilz, Hellweg, Harteis, Rückert, and Schneider, the concept of the 'global digital twins of humans' is defined and compared with the established concept of the 'digital twins of machines'. Possible data sources for the digital twins of humans, which are divided into genetic and biographical data sources, are presented.

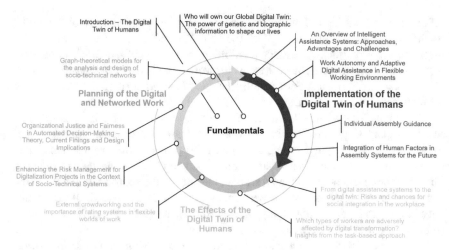

Fig. 1 Overview of the book

In the context of **Planning Digital and Networked Work**, the question of which design criteria must be taken into account in the design of the digital world of work when using the digital twins of humans and how this can be done is raised. The early consideration of risks arising from the use of the digital twins of humans and the appropriate modelling of the entire socio-technical system are crucial.

In the chapter *Enhancing the Risk Management for Digitalisation Projects in the Context of Socio-Technical Systems* by Menzefricke, Koldewey, and Dumitrescu, an extension of existing risk management approaches that includes a socio-technical approach is presented. When introducing a digital transformation, such as the digital twins of humans, it is crucial to take a socio-technical view and also to consider social and organisational risks at an early stage.

In the chapter *Justice and Fairness Perceptions in Automated Decision-Making—Current Findings and Design Implications* by Hellwig and Maier, the effects of automated decisions on perceptions of justice are examined. For this purpose, a comprehensive literature review on the perceptions of justice and fairness in automated decision-making is performed. On the one hand, a comparison between automated and human decision-making is made, and on the other hand, design and personality characteristics that influence the perception of justice and fairness are investigated.

In the chapter *Graph-Theoretical Models for the Analysis and Design of Socio-Technical Networks* by Cappello and Steffen, procedures for planning and analysing socio-technical systems and thus also the digitalised world of work are presented. Signed graphs, as well as coloured signed graphs, for modelling these systems are presented. By using the presented procedures, global statements can be made about the socio-technical networks of the digitalised world of work. The mathematical foundations and procedures for using these approaches are described in this chapter.

In the context of the **Implementation of the Digital Twins of Humans**, the question of how the digital world of work can be created with the use of the digital twins of humans is considered. For this purpose, technical implementations of the digital twins of humans are presented. With the help of digital twins, various automated decision support systems are made possible in the context of industrial production; for example, adaptive assistance systems for worker guidance in an industrial assembly context can be created. Furthermore, the role of the human digital twin in relation to work autonomy is shown.

In the chapter *Adaptive Assistance Systems: Approaches, Benefits, and Risks* by Buchholz and Kopp, an overview of different adaptive assistance systems, types of assistance, and applied strategies is given. To provide effective and acceptable adaptive assistance, technical systems must first keep track of the state of the task, the environment, and the user, and, second, they must take appropriate actions at the right time in order to improve the performance criteria of interest. To illustrate this, an architecture for an adaptive assistance system is presented. This assistance system supports the user in performing monitoring tasks.

The chapter *Work Autonomy and Adaptive Digital Assistance in Flexible Working Environments*, by Gensler, Oestreich, Abendroth, Wrede, and Wrede, creates a uniform understanding of the work autonomy of humans and technical systems and shows how adaptive assistance systems can support this autonomy. Based on the analysis of sociological and technical autonomy, nine possible scenarios are derived that illustrate how human work autonomy and system autonomy relate to each other. On the basis of these scenarios, guidelines for the development of adaptive assistance systems that increase the work autonomy of employees are presented.

In the chapter *Individual Assembly Guidance* by Pöhler and Gräßler, the technical implementation of an adaptive assistance system for an assembly station in a laboratory environment is described. The assistance system learns about the worker during the assembly process and thus develops the digital twin of the assembler. This forms the basis for the adaptive support of the worker.

In the chapter *Integration of Human Factors for Assembly Systems of the Future* by Roesmann and Gräßler, a framework for how the digital twins of humans can support the scheduling of assembly processes in cyber-physical assembly systems is presented. The presented framework extends the existing procedures for scheduling assembly processes using a human-centred perspective. This contribution focuses on the model-based description of the workers through the use of human factors.

In the context of **The Effects of the Digital Twins of Humans**, the following question can be asked: What effects will the use of the digital twins of humans have on the working world of the future? To answer this question, the results of occupational and sociological studies are presented. These findings are crucial for both planning for and implementing the digital twins of humans.

In the chapter *From Computer-Assisted Work to the Digital Twins of Humans: Risks and Chances for Social Integration in the Workplace* by Brunsmeier, Diewald, and Reimannn, there is a discussion of how the design, implementation, and individual appropriation of computer-assisted work can shape working relationships. This research focuses on the role of the digital twins of humans in social integration in the workplace.

In the chapter *Which Types of Workers are Adversely Affected by Digital Transformation? Insights from the Task-based Approach* by Hellweg and Schneider, the results of a task-based approach are presented. These results show which skills and types of jobs are particularly affected by digital transformation. Furthermore, statements are made on specific training and retraining needs.

In the chapter *Digital Twins in Flexible Online Work: Crowdworkers on German-language Platforms* by Hemsen, Reimann, and Schneider, a new form of work called crowdworking is described. Empirical findings on the characteristics of crowdworking platforms in Germany are presented. The opportunities and risks of this form of work are discussed against the background of the digital twins of humans.

As the editors of this book, we are convinced that the development of the digital twins of humans will be a central element of the optimisation and design of complex socio-technical systems in digitised work. The following chapters are intended to contribute to planning for, designing, and using human digital twins.

Acknowledgements Iris Gräßler, Günter W. Maier, Eckhard Steffen, and Daniel Roesmann are members of the research programme 'Design of Flexible Work Environments—Human-Centric Use of Cyber-Physical Systems in Industry 4.0', which is supported by the North Rhine-Westphalian funding scheme 'Forschungskolleg'.

References

Boschert, S., & Rosen, R. (2016). Digital twin—The simulation aspect. In Hehenberger P (Ed.) *Mechatronic futures* (pp. 59 –74). Springer. https://doi.org/10.1007/978-3-319-32156-1_5

Buchholz, V., & Kopp, S. (2020). Towards an adaptive assistance system for monitoring tasks: Assessing mental workload using eye-tracking and performance measures. In *International conference on human-machine systems* (pp. 1–6). Piscataway, NJ: IEEE. https://doi.org/10.1109/ICHMS49158.2020.9209435

DHL. (2019). DHL Trend-Report – Einsatz von digitalen Zwillingen verbessert Logistik-abläufe deutlich. www.dpdhl.com/content/dam/dpdhl/de/media-relations/press-releases/2019/pm-trend-report-digital-twins-20190627.pdf

Engels, G. (2020). Der digitale Fußabdruck, Schatten oder Zwilling von Maschinen und Menschen. *Zeitschrift für Angewandte Organisationspsychologie, 51*(3), 363–370. https://doi.org/10.1007/s11612-020-00527-9

Gartner. (2018). Gartner top 10 strategic technology trends for 2019. https://www.gartner.com/smarterwithgartner/gartner-top-10-strategic-technology-trends-for-2019

Gensler, E., & Abendroth, A. K. (2021). Verstärkt algorithmische Arbeitssteuerung Ungleichheiten in Arbeitsautonomie? *Soziale Welt, 72*(4), 514–550. https://doi.org/10.5771/0038-6073-2021-4-514

Gräßler, I. (2003). Impacts of information management on customized vehicles and after-sales services. *International Journal of Computer Integrated Manufacturing, 16*(7–8), 566–570. https://doi.org/10.1080/0951192031000115714

Gräßler, I., Pöhler, A. (2017). Integration of a digital twin as human representation in a scheduling procedure of a cyber-physical production system. In *International conference on industrial engineering & engineering management* (pp 289–293). Piscataway, NJ: IEEE. https://doi.org/10.1109/IEEM.2017.8289898

Gräßler, I., & Pöhler, A. (2019). Human-centric design of cyber-physical production systems. *CIRP Design, 84*, 251–256. https://doi.org/10.1016/j.procir.2019.04.199

Gräßler, I., & Pöhler, A. (2020). Produktentstehung im Zeitalter von Industrie 4.0. In Maier GW, Engels G, Steffen E (eds) *Handbuch Gestaltung digitaler und vernetzter Arbeitswelten* (pp. 383–403). Berlin, Heidelberg: Springer. https://doi.org/10.1007/978-3-662-52979-9_23

Gräßler, I., Roesmann, D., & Pottebaum, J. (2020a). Model based integration of human characteristics in production systems: A literature survey. In *CIRP conference on intelligent computation in manufacturing engineering* (pp. 57–62). Elsevier.

Gräßler, I., Roesmann, D., & Pottebaum, J. (2020b). Traceable learning effects by use of digital adaptive assistance in production. In *Conference on learning factories* (pp. 479–484). Elsevier.

Gräßler, I., Roesmann, D., Cappello, C., & Steffen, E. (2021). Skill-based worker assignment in a manual assembly line. *CIRP Design, 100*, 433–438. https://doi.org/10.1016/j.procir.2021.05.100

Gräßler, I., Roesmann, D., Hillebrand, S., & Pottebaum, J. (2022). Information model for hybrid prototyping in design reviews of assembly stations. In *CIRP conference on intelligent computation in manufacturing engineering* (pp. 489–494). https://doi.org/10.1016/j.procir.2022.09.054

Grievers, M. (2014). Digital twin: Manufacturing excellence through virtual factory replication.

Grosse, E. H., Calzavara, M., Glock, C. H., & Sgarbossa, F. (2017). Incorporating human factors into decision support models for production and logistics: Current state of research. *IFAC-PapersOnLine, 50*(1), 6900–6905. https://doi.org/10.1016/j.ifacol.2017.08.1214

Hellwig, P., Buchholz, V., Kopp, S., & Maier, G. W. (2023). Let the user have a say - Voice in automated decision-making. *Computers in Human Behavior, 138*, 107446. https://doi.org/10.1016/j.chb.2022.107446

Kauffeld, S., & Maier, G. W. (2020). Schöne digitale Arbeitswelt - Chancen. *Risiken und Herausforderungen. Gruppe Interaktion Organisation Zeitschrift für Angewandte Organisationspsychologie (GIO)*.

Liu, M., Fang, S., Dong, H., & Xu, C. (2021). Review of digital twin about concepts, technologies, and industrial applications. *Journal of Manufacturing Systems, 58*, 346–361. https://doi.org/10.1016/j.jmsy.2020.06.017

Macchi, M., Roda, I., Negri, E., & Fumagalli, L. (2018). Exploring the role of digital twin for asset lifecycle management. *IFAC-PapersOnLine, 51*(11), 790–795. https://doi.org/10.1016/j.ifacol.2018.08.415

Miller, M. E., & Spatz, E. (2022). A unified view of a human digital twin. *Human-Intelligent Systems Integration, 4*(1–2), 23–33. https://doi.org/10.1007/s42454-022-00041-x

Mlekus, L., Lehmann, J., & Maier, G. W. (2022). New work situations call for familiar work design methods: Effects of task rotation and how they are mediated in a technology-supported workplace. *Frontiers in Psychology, 13*. https://doi.org/10.3389/fpsyg.2022.935952

Oestreich, H., Töniges, T., Wojtynek, M., & Wrede, S. (2019). Interactive learning of assembly processes using digital assistance. *Procedia Manufacturing, 31*, 14–19. https://doi.org/10.1016/j.promfg.2019.03.003

Ötting, S. K., & Maier, G. W. (2018). The importance of procedural justice in human-machine interactions: Intelligent systems as new decision agents in organizations. *Computers in Human Behavior, 89*, 27–39. https://doi.org/10.1016/j.chb.2018.07.022

Shengli, W. (2021). Is human digital twin possible? *Computer Methods and Programs in Biomedicine Update, 1*, 100014. https://doi.org/10.1016/j.cmpbup.2021.100014

Stark, R., Anderl, R., Thoben, K. D., & Wartzack, S. (2020). WiGeP-Positionspapier: Digitaler Zwilling. *ZWF Zeitschrift für wirtschaftlichen Fabrikbetrieb, 115*, 47–50. https://doi.org/10.3139/104.112311

Who Will Own Our Global Digital Twin: The Power of Genetic and Biographic Information to Shape Our Lives

Sarah Pilz, Talea Hellweg, Christian Harteis, Ulrich Rückert, and Martin Schneider

Abstract Today, it is possible to collect and connect large amounts of digital data from various sources and life domains. This chapter examines the potential and the risks of this development from an interdisciplinary perspective. It defines the 'global digital twin' of a human being as the sum of all digitally stored information and predictive knowledge about a person. It points out that, compared to the digital twin of a machine, the human global digital twin is far more complex because it comprises the genetic code and the biographic code of a person. The genetic code contains not only a simple 'construction plan' but also hereditary information, in a form that is difficult to read. The biographic code contains all other information that can be assembled about a person, which is obtained via data from cameras, microphones, or other sensors, as well as general personal information. When the growing wealth of information concerning the genetic code and the biographical code is properly utilised, insights from biology and the behavioural sciences may be used to predict personal events such as health problems, job resignations, or even crimes. Because our own interests and those of private firms are partly in conflict over the use of this powerful knowledge, it is still unclear whether the global digital twins of humans will become a liberating or disciplining force for citizens. On the one hand, human beings are not machines: They are aware of their digital twin and therefore are able to influence it throughout their lives. Because of their free will, human beings are in

S. Pilz and T. Hellweg contributed equally to this work.

S. Pilz (✉) · U. Rückert
CITEC – Bielefeld University, Inspiration 1, 33619 Bielefeld, Germany
e-mail: spilz@techfak.uni-bielefeld.de

U. Rückert
e-mail: rueckert@techfak.uni-bielefeld.de

T. Hellweg · C. Harteis · M. Schneider
Paderborn University, Warburger Str. 100, 33098 Paderborn, Germany
e-mail: talea.hellweg@uni-paderborn.de

C. Harteis
e-mail: christian.harteis@uni-paderborn.de

M. Schneider
e-mail: martin.schneider@uni-paderborn.de

general difficult to predict. Dystopias of full control over individual behaviour are therefore unlikely to materialise. On the other hand, private firms are beginning to take advantage of the available digital twins of humans by monopolising data access and by commercialising predictive knowledge. This is problematic because, unlike machines, human beings cannot only benefit from but also suffer due to their digital twins as they attempt to shape their own lives. We illustrate these issues with some examples and arrive at two conclusions: It is in the public interest for people to be granted more property rights over their personal global digital twins, and publicly funded research needs to become more interdisciplinary, much like private firms that have already begun to perform interdisciplinary research.

Keywords Digital twin · Genetic data · Biographic data · Global · Free will · Common pool resource

1 Introduction

Human beings of the 21st century produce online data almost continuously. We choose a radio program or playlist in the morning, check in to work on the computer, have our lunch delivered, are filmed when we park our bike in town, pay for a newspaper and a coffee by credit card, document the distance we run and the calories we burn while jogging, switch on the lights when we enter our smart home, browse travel destinations after dinner, fill out a customer questionnaire, and watch our favourite film before we set an alarm on our mobile phone. On certain days, we leave digital traces with a physician or a hospital, with career websites or social networks, with political parties or interest groups, with social security administrations, and with our insurance company.

The huge amount of information about a person in a specific domain, such as healthcare, education, or work, can be used to construct a human digital twin (DT) (see Barricelli et al. (2020), Berisha-Gawlowski et al. (2021), and Graessler and Poehler (2018)). While the DT concept is still predominantly applied to non-living objects such as machines (Barricelli et al., 2019), the idea has been instructive for understanding how data can be used to produce an accurate picture of a person's state of health, career success, or skill profile (Bruynseels et al., 2018; Graessler & Poehler, 2018). Domain-specific DTs of humans have the potential to improve our lives, for example, by preventing illnesses or finding us better jobs. Today, we are not aware of all the information available about us, and in many cases private firms own the data that constitute our DTs. Therefore, the potential benefits of DTs may not be apparent to us as citizens. There is even some danger that DTs will be applied to discipline or manipulate us. In Industry 4.0, interconnectedness is considered a productive feature on which the digital twins of machines can build.

Will the benefits to citizens outweigh the potential drawbacks as more networked information about citizens becomes available on the internet? This chapter attempts to provide answers by making three contributions. First, the new concept of the 'global digital twin' is introduced. It is defined as the sum of all the digitally stored information and predictive knowledge concerning an individual. Rather than being restricted to a particular application, it includes all domain-specific DTs. Second, the concept of the global digital twin of a human being is unfolded by distinguishing two categories of information, namely the biographic code and the genetic code. Linking these codes is especially useful for creating predictive knowledge concerning important issues such as job choices, health risks, or consumer preferences. The complexity of the codes forestalls the possibility that the global digital twin will be exploited to gain full control over citizens. Third, it is argued that, from an economics perspective, the knowledge that makes up the global digital twin is a common pool resource, and this has important policy implications.

This chapter builds on various streams of research and is a highly interdisciplinary essay. Section 2, building on various scientific disciplines, derives a definition of the global digital twin of a human being, juxtaposing it with the concept of the DTs of machines. Section 3 discusses the concept in detail, summarises how data are collected and integrated, and, in a philosophical aside, illustrates how free will must be considered to understand the function and limits of the global digital twin. We conclude that data and knowledge are becoming key resources. Section 4, therefore, sheds light on the economic mechanisms that govern how data and knowledge are created and which incentive problems are inherent in this process. Section 5 concludes by sketching the implications for public policy and the organisation of publicly funded research.

2 Defining the Global Digital Twin of a Human Being

While there appears to be a common basic understanding of the digital twin in general, this is not true for the human digital twin. For this essay, we first researched the origins of the term 'digital twin' in the field of machines, where it originated (see Sect. 2.1); this was followed by extensive literature research on the digital twins of humans (see Sect. 2.2). It became evident that the idea of a human digital twin is used in various disciplines and is understood slightly differently depending on the application context. In order to describe a general human digital twin, we first define it as comprising all digitally stored data about a person and call it the 'global digital twin'. We use the term 'global' here to mean all-encompassing, since we refer to the generation of a digital twin for each person that includes all possible collectible data. These data include both raw data and analytical findings (see Sect. 2.3.4) generated by different actors, such as internet firms, employers, health institutions, and public agencies. Therefore, the global digital twin contains a wide variety of information that is collected, stored, and used by different institutions. Since the global digital twin is a theoretical construct and not all data are available in this bundled form, in reality, many local or so-called sub-twins contain partial data and are stored and analysed by different actors. While the digital twin of an employee in a company, for example, consists of information about the person's skills, the online retailer

Amazon collects information about the person's purchasing behaviour. Both types of information are part of a person's global digital twin, which is made up of these sub-twins. Digital twins of individual persons can also be combined. These joint twins, consisting of a multitude of digital twins of different individuals, can be used, for example, to create reliable databases for research. This is done, for example, by Jacqueline Alderson and her colleagues, who combine personal data such as 3D scans, PET scans, X-rays, images, functional data, etc., from many people into a digital twin. The resulting 'digital athlete' can be used, for example, to derive the strain on a person's knee during exercise from 2D images without having to perform all these scans on the person in question (Alderson & Johnson, 2016; Johnson et al., 2018).

We then argue that the data can be divided into two large components, namely the genetic code and the biographical code (see Sect. 2.3). In our definition, the genetic code contains the DNA and hereditary information; the biographical code contains all other information that can be assembled about a person throughout their life. This distinction is based on the use of the term 'digital twin' in genomic research and medicine, as well as in studies of biographical events. In addition, the distinction results from fundamental differences between the two forms of data. The two categories can also be viewed as counterparts to the two basic elements that are distinguished for the digital twins of products and machines. The genetic code represents the basic blueprint of a human being and thus has properties that are similar to the building plan of a machine. The biographical code contains information about behaviour and external circumstances and corresponds to the life-cycle data of a machine. The rough data classifications of digital twins and human digital twins are thus comparable.

2.1 Digital Twins of Machines

To provide a general background for our analysis, we reviewed the different definitions of the digital twin of a machine. Terminologies differed: While the basic idea for twin models has certainly existed for some time, the basic concept of the digital twins of machines has remained constant since the early 2000s (Grieves, 2015). Michael Grieves introduced conceptual ideas for product life-cycle management. This is often seen as the first concrete application of the digital twin concept in manufacturing.

In 2015, Grieves specified that the digital twin 'contains three main parts: (a) physical products in Real Space, (b) virtual products in Virtual Space, and (c) the connections of data and information that ties the virtual and real products together' (Grieves, 2015). Another frequently cited definition of a digital twin was introduced by NASA and described a digital twin as 'an integrated multiphysics, multiscale, probabilistic simulation of an as-built vehicle or system that uses the best available physical models, sensor updates, fleet history, etc., to mirror the life of its corresponding flying twin' (Glaessgen & Stargel, 2012). Since then, a variety of definitions of the digital twin have emerged, depending on the context (for an overview, see Barricelli et al. (2019), Fuller et al. (2020), Lim et al. (2020), and Jones et al. (2020)). Barricelli et al. reviewed 75 papers about digital twins, less than half of which included a definition. From the aforementioned papers, they listed 29 different definitions, grouped according to six key points: 'Integrated system', 'Clone, counterpart', 'Ties, link', 'Description, construct, information', 'Simulation, test, prediction', and 'Virtual, mirror, replica' (Barricelli et al., 2019). Fuller stated that 'the Digital Twin is defined extensively but is best described as the effortless integration of data between a physical and virtual machine in either direction' (Fuller et al., 2020). Overall, a clear, uniform definition of the digital twin does not exist, but there appears to be a common understanding of what the digital twin of a machine or physical object is.

2.2   Literature: Digital Twins of Humans

It is difficult to apply all defining characteristics of the digital twins of machines to a human digital twin, so some adaptations are necessary (El Saddik, 2018). Human beings and technology are fundamentally different. Human behaviour is much more complex and difficult to model than that of machines. Considerably more influencing factors must be taken into account, and data collection via sensors is not possible in the same way that it is when the digital twin of a machine is created. Additionally, a continuous connection to the network/internet to update the digital twin of a human is not as simple as it is for machines. To explore how the human digital twin is applied in different fields and how it is defined, we searched relevant databases (EBSCO, Google Scholar, IEEE) for literature on the human digital twin. The findings are summarised in Table 1. This list of definitions does not claim to be exhaustive, but it is intended to provide central aspects of the consideration of the human digital twin in various research areas. So far, the idea of a human digital twin has been used mainly in the medical field, with terms such as the digital patient, physical twin, or DTs of organs (Bruynseels et al., 2018). However, it is also increasingly used in other areas, such as production planning and control, marketing, management, and robotics (see Table 1). In these areas, different types of human digital twins are emerging: the digital citizen (Mossberger et al., 2007; Hintz et al., 2018), the digital athlete (Alderson & Johnson, 2016; Barricelli et al., 2020), and the digital twin for employees (Graessler & Poehler, 2018). Various attempts have been made to create a common understanding of the human digital twin. However, these definitions are mostly contextual; they depend on the specific application and field of expertise. Hence, different understandings of what we consider digital sub-twins currently exist side by side, and there is no general definition (Engels, 2020; Fuller et al., 2020).

Table 1 Definitions of the human digital twin in the literature

Field	Source	Example definitions
Medicine/healthcare	Alderson and Johnson (2016), Bruynseels et al. (2018), Chakshu et al. (2020), Barricelli et al. (2020), Erol et al. (2020), Matusiewicz et al. (2019), Croatti et al. (2020), Rodriguez (2020), Voigt et al. (2021), Terpsma and Hovey (2020), Shengli (2021), Gámez Díaz et al. (2020), Popa et al. (2021), Jimenez et al. (2020), Bagaria et al. (2020), Laamarti et al. (2020)	'When applied to persons, Digital Twins are an emerging technology that builds on in silico representations of an individual that dynamically reflect molecular status, physiological status and life style over time' (Bruynseels et al., 2018). 'Human DTs differ from DTs developed and used in Industry 4.0 because they are not continuously connected to their physical twin. In this case, users and experts are meant to continuously update the DT with the input of medical digital data describing the health condition of the physical twin' (Barricelli et al., 2020). '[...] redefinition of the digital twin from the digital replica of a physical asset to the replica of a living or nonliving entity [...]' (Laamarti et al., 2020)
Production planning and control	Graessler and Poehler (2018), Yigitbas et al. (2021), Petzoldt et al. (2020), Baskaran et al. (2019), Joseph et al. (2020), Kemény et al. (2018), Nikolakis et al. (2019), Hafez (2020)	'The idea of a digital twin is to digitally emulate real behavior [...]. Adapted to a human digital twin, it serves as linkage between skills and demands of employees and the technical system. This idea of digital twins is based on a software solution, which includes digital representations of employees in computational decision-making processes. Human characteristics like behavior, skills and preferences of the employee are digitalized. The digital twin of employees uses this information in order to negotiate with technical devices according to the employee's settings. The digital twin is capable to process user feedback and recorded patterns' (Graessler & Poehler, 2018)

(continued)

Table 1 (continued)

Field	Source	Example definitions
Marketing	Truby and Brown (2021), Sun et al. (2021)	'The digital twins in this framework have several characteristics. (1) Sensing and actuating: Real twins can be equipped with sensors to replicate their senses, namely, sight, hearing, taste, smell, and touch, using appropriate actuators depending on application needs. In online social networks, the posting content and liking behavior of users can be recorded to sense and understand users' personality' (Sun et al., 2021)
Management	Berisha-Gawlowski et al. (2021), Dorrer (2020), Kesti (2021), Dorrer (2020), Mossberger et al. (2007), Hintz et al. (2018), Heinke (2021), Hafez (2020), Zibuschka et al. (2020), Gomerova et al. (2021)	'The prospect of a human digital twin (HDT) is to develop a digital representative of a human user, taking into account the user's goals and preferences, what the user knows, and what the user does while interacting with services, assistants, and the Internet of Things (IoT) environment' (Heinke, 2021)
Robotics	Bryndin (2019a), Bryndin (2019b), Trobinger et al. (2021)	'Human digital twins provide services in the social sphere and in space. Training of digital twins in professional competences is carried out on the basis of communicative associative logic of technological thinking by cognitive methods. Cognitive psychology experts investigating the effectiveness of machine learning techniques offer a new approach that allows artificial intelligence and cognitive psychology to be combined. This approach provides pre-preparation of neural networks from accumulated data using existing behaviors' (Bryndin, 2019b)

2.3 Genetic and Biographic Codes

Our definition of the global human digital twin includes all types of information, which we divide into genetic and biographic data. We speak more precisely of the genetic and the biographic 'codes' because the data must be not only recorded but also

Fig. 1 A global digital twin
consists of genetic and
biographic data

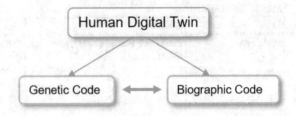

processed and interpreted. The data stem from a variety of sources in many different
fields. Sometimes, data production is based on scientific insights (blood pressure),
and sometimes the information is a by-product of a certain process (social media
entries). Almost all data are important because they allow the analysis and prediction
of behaviour (such as consumption) or states (such as health). However, making
reliable predictions hinges on the insights, theories, and accumulated knowledge
of various, highly diverse scientific disciplines. This is further complicated by the
fact that the two codes influence each other (see Sect. 2.3.3). Therefore, the important
point that our definition of the global human DT suggests is that advances in predictive
knowledge are likely to come from integrating biographic and genetic data and, by
implication, from highly interdisciplinary research (see Fig. 1).

2.3.1 The Genetic Code

The basic blueprint of a human is stored in their genetic data. More precisely, the
term 'genetic code' refers to the specific DNA sequence of a person. Sequencing
determines the order of the four different nucleotides[1] in the sequence. The complete
DNA sequence of a human contains more than 3 billion base pairs. Although DNA
was discovered in the 1860s (Dahm, 2008), its general structure was not identified
until the 1950s (Watson & Crick, 1953). The first complete human DNA sequence
was published in 2003. It was determined over a period of 13 years as part of the
large-scale Human Genome Project. The sequence was not that of a single person;
samples from several volunteers were analysed. It should be noted that the genomes
of any two persons are about 99.9% identical (NIH, 2018). Overall, the project
cost 2.7 billion US dollars (NIH, 2020). Today, the individual DNA sequence of a
person can easily be determined by new sequencing technologies, and this process
is becoming increasingly affordable. With modern sequencing methods, a complete
human genome can be determined for less than 1,000 US dollars (Wetterstrand,
2020).

The far more complicated task is to find out what the determined DNA sequence
means, i.e., what information it contains. This is difficult to determine because the
DNA sequence is a 3-billion-character-long code composed of the letters A, C, G,

---

[1] These nucleotides are distinguished by their different bases, adenine (A), cytosine (C), guanine
(G), and thymine (T).

and T. In other words, if one were to write down the DNA sequence of a human being in the form of a series of books, it would take over 4,500 books of 300 pages each with 2,295 characters per page to print it—thousands of books of which a human being would not understand a word. If one wants to understand the information contained in DNA, one must first consider that three of the nucleotides/letters are a code for one of 20 amino acids or the beginning/end of an amino acid sequence. From here on, it becomes more and more complicated to understand what exactly is happening. These translated amino acid sequences (e.g., proteins) perform a multitude of different tasks in the human body. Which amino acid sequences are read and built and when this occurs depends on many factors. Even determining what the result will look like is very complicated. Additionally, the function of each of these sequences has been the subject of intensive research for many years.[2] There are several large research projects that followed the Human Genome Project and aimed to fully understand the human genome. For example, the ENCODE (ENCylopedia Of DNA Elements) project, which has been carrying out research since 2003, studies which genes humans possess and their specific functions (Consortium et al., 2012; Davis et al., 2018). The next step is not only to sequence and understand individual people's genomes but also to use this knowledge to identify, for example, perfectly matching (personalised) therapies for diseases. The opportunities and risks arising from combining these genetic data with biographical data are explained in Sect. 2.3.3.

2.3.2 The Biographic Code

Biographic data contains, in general, all data except genetic data about a person that can be measured. Almost 80 years ago, Vannevar Bush introduced the idea of the Memex (Bush, 1945), a hypothetical device for an individual library with fast and flexible access to everything a person has ever read, including communications. Bush described it as 'an enlarged intimate supplement to his memory'. In 2001, Microsoft established the project MyLifeBits (Microsoft Research, 2017), intending to fulfil the vision of the Memex. In addition to textual content, audio recordings, videos, and photos, among other things, were stored. The researcher Gordon Bell even wore a camera that automatically took pictures during the day. Everything was stored digitally. At the same time, software to annotate, store, and search these large amounts of data was developed. In 2016, Bell said in an interview that this attempt to perform lifelong storage was too early (Elgan, 2016). He assumes that in the future, so-called lifelogging will become popular again. This will require cheaper memory, longer battery life, and better artificial intelligence applications. For now, he states that smartphones are a version of Bush's Memex. Overall, there is a wide range of data that can be stored about an individual, from general personal information, such as their name and date of birth, to location data, personal preferences, their health status, and much more (see also Table 2).

---

[2] Note: This is a highly simplified explanation.

Which items and types of data should be included when we want to predict behaviour? Predicting a person's health status, job choice, election behaviour, or consumption is difficult to do. Much like genetic information, biographic information is also a code. In addition to choosing among data pools, one can also indirectly derive information by combining directly collected raw data. For example, if you take a person's debit card statements, you can see directly how much money they spent in a particular store. However, these statements also include the information that this person visited the store on a particular day. Many of these data could be used to record positional data or to derive preferences for certain types of stores (discount stores, supermarkets, boutiques, etc.). From this, the approximate place of residence, income, whether the person has children, and so on, could be inferred. In addition, data networks can be developed that compile different types of information and map relationships (see Sect. 3.2 for more information).

2.3.3 Combining the Codes

The biographical code and the genetic code influence each other. As the genetic code represents the basic building blocks of human beings, this code influences people's biographical information. It does not just determine how a person looks (e.g., their height, eye/hair/skin colour, basic body shape, etc.); people can also have a genetic predisposition toward, e.g., certain behaviours, abilities, and diseases, which may also be visible in the biographical code. In the 19th century, before the discovery of DNA, the scientific literature discussed whether acquired characteristics are inherited from the previous generation. The prevailing opinion was that this is not the case, e.g., the children of parents who play sports are not necessarily born athletes. Observed effects were attributed to education and natural selection. Recent findings in the field of epigenetics suggest that the biographical code also influences the genetic code (Egger et al., 2004; Suzuki & Bird, 2008). Epigenetics shows that human behaviour (e.g., nutrition, trauma, etc.) influences the reading of DNA. Although the nucleotide sequence of DNA remains constant throughout a person's lifetime (no mutations or recombinations), which genes are actually read and which are not read can change and frequently does. This happens as these genes are activated or deactivated by enzymes. Of particular interest here are studies of identical twins, who have identical genomes (Fraga et al., 2005). There are cases in which the twins have a genetic predisposition for a certain disease, but only one twin becomes sick; the other stays healthy.

These changes in how DNA is read not only affect the individual but also may have an impact on the next generation. A common example of epigenetics is the study of the effects of mothers' and fathers' eating behaviours on their offspring. In an experiment with laboratory mice, oocytes and sperm from mice that were fed a high-calorie and high-fat, normal, or low-calorie diet for several weeks were implanted into healthy surrogate mothers, and the offspring were again fed a high-calorie and high-fat diet (Huypens et al., 2016). Researchers were able to show that epigenetic information on eating behaviour is passed on to the next generation. This leads to the result that the offspring of mice that ate an unhealthy diet (high-calorie and high-fat) suffered more frequently from obesity and problems with blood glucose regulation. In addition, another example shows that environmental influences can also have a direct impact on the behaviour of individuals via epigenetic mechanisms. Research findings suggest that long-term epigenetic changes occur in the brains of adults with a history of childhood trauma (Thumfart et al., 2021; Labonte et al., 2012). These changes may, for example, cause some mental illnesses that occur in old age (Thumfart et al., 2021).

The two examples illustrate that epigenetics represents a link between environmental influences, i.e., biographical data, and genes. The data types (biographical and genetic) are therefore linked. Thus, if both types of data are considered together, this can lead to more accurate and reliable information. This obviously entails opportunities as well as risks (see also Sects. 3 and 4).

2.3.4 Data Layout of the Digital Twin

Figure 2 shows the data layout of the (raw) data of a human digital twin, illustrating that part of the data can be influenced by both the biographic and genetic codes. In addition, the digital twin stores technical data about interfaces, memory structure, connected sensors, etc. While the global digital twin is defined as containing all potentially measurable types of data, there will consequently be some data that cannot be stored or measured. This includes, in particular, the 'free will' of a person, which is therefore represented outside the digital twin (see Sect. 3.3). The diagram shows

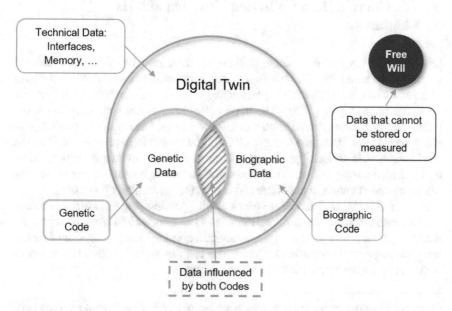

Fig. 2 Data layout of the human digital twin

only one possible layout of the first stage of data storage in the digital twin. This stage consists of raw data, e.g., the outputs from sensors. At this stage, the data have not yet been interpreted in any way, and it is unknown what information is contained indirectly in the data or could be created by linking data. Therefore, these raw data represent the genetic and biographic codes stored in the digital twin. In further stages, which are not shown in the diagram, the results emerging from the evaluation and interpretation of the stored raw data are also included in the global digital twin. Sub-twins only contain delimited parts of the data and may only contain interpreted (rather than raw) data about a person. Which data are contained in a sub-twin and how they are represented depends on the application. Both codes, genetic and biographic, and their various interactions need to be deciphered. Therefore, the (global) digital twin also includes information that transforms data into knowledge, which can be used to predict behaviour. This is the type of knowledge that biology and the social sciences are working on; it is less developed than the pure information in the codes. Some questions of interest, among others, are the following: How likely is it that people will remain healthy and productive workers, given their genetic disposition, their exercise behaviour, and their leisure habits? Can a certain diet alleviate some health risks? How likely is it that people who are extremely neurotic will fail to cooperate with their colleagues? How likely are they to renege on their debts? This type of information is often gained from questionnaire surveys on people's opinions and personality traits.

3   The Current Data Collection Situation and Its Limitations

Due to its wide range of applications and benefits, the human digital twin is considered an emerging technology. This is also reflected in the classification of the digital twin of a person[3] as one of the driving technologies in the Gartner hype cycle for new emerging technologies (2020). The hype cycles represent the maturity, adoption, public attention, and societal applications of emerging technologies. According to the hype cycle of 2020 (see Gartner (2020)), the digital twin of a person is in the 'early adopters investigate' phase and will move over time to the next phase, which is 'mass media hype begins'. This shows that the research field is expected to gain relevance and attention in the future and is therefore particularly interesting.

In the following, we will therefore look at what possible data from various sources can be contained in the human digital twin. In this section, we will also look at how the data, e.g., from well-known companies, are processed to create an integrated model for generating predictions about people. The last part of this section is devoted to what cannot be predicted: the free will of a person.

---

[3] Gartner summarises several technologies that are part of a leading trend called the 'digital me' (see Gartner (2020)). It contains, e.g., the digital twin of a person, citizen twins, health passports, and bidirectional brain-machine interfaces.

3.1 Data Sources

There are a variety of possible data sources and types of data that can be stored about a person. For a coarse categorisation, we refer to the European General Data Protection Regulation (GDPR).

The GDPR defines personal data in Article 4 as 'any information relating to an identified or identifiable natural person ("data subject"); an identifiable natural person is one who can be identified, directly or indirectly, in particular by reference to an identifier such as a name, an identification number, location data, an online identifier or to one or more factors specific to the physical, physiological, genetic, mental, economic, cultural or social identity of that natural person' (GDPR, 2016).

The following is also stated in article 9: 'Processing of personal data revealing racial or ethnic origin, political opinions, religious or philosophical beliefs, or trade union membership, and the processing of genetic data, biometric data for the purpose of uniquely identifying a natural person, data concerning health or data concerning a natural person's sex life or sexual orientation shall be prohibited' (GDPR, 2016).

Table 2 shows some examples of which data are included in these categories. While 'raw genetic data' is only a single category in the table, genetic data, in general, influence much of the data in the other categories. For example, many physical, physiological, and mental characteristics, as well as health data, are strongly influenced by the genetics of a person. There are also somewhat strong effects on a person's cultural and social identity and on many other categories. The table is not a complete list of all possible types of data but gives an idea of what data about a person can be stored. It should be noted how little data are sometimes needed to uniquely identify a person. According to a study from 2000, for example, 87% of U.S. citizens can be uniquely identified if their gender, date of birth, and postal code are known (Sweeney, 2000).

3.2 Data Integration and Prediction

The previous subsection illustrated that a variety of data are in principle available for predicting human behaviour or states such as health. As Sect. 2.3.3 argued, the most dramatic progress in building such predictive models will probably result from combining the biographic and genetic codes.

Such integration has been achieved to some extent in the area of healthcare. Personalised medicine has been made possible by analysing big health data on individuals. Not surprisingly, it is here that the idea of a human digital twin was suggested first (Bruynseels et al., 2018). Personalised medicine relies on an image of each individual based on genetic data, physiological data, and lifestyle decisions: In other words, it relies on a combination of the genetic and biographic codes. Fine-grained personalisation does not work, however, unless the data for millions of patients become available. These big data, in combination with appropriate theories, permit scientists

Table 2 Data categories from GDPR and examples (given by the authors)

Category	Examples
General personal data	Name
	Date of birth
	City of birth
	Residence
	Education
Identification numbers	Social insurance numbers
	Tax IDs
	Identity card numbers
Bank data	Account balance
	Credit rating
Location data	GPS position
	Place of use of debit card
Online identifier	Cookies
	IP addresses
Physical/physiological characteristics	Gender
	Biometric data
	Eye and hair colour
	Height
	Clothing size
	Athletic performance
Raw genetic data	DNA sequence
	Inherited traits
	Inherited diseases
Mental characteristics	Mood
	Mental illnesses
	Personality
	Cognitive measures
	Origin of drive
Economic characteristics	Labour force status
	Wage
	Tax class
Cultural identity	Society
	Cultural milieu
	Subculture
Social identity	(Dis)abilities
	Social class
	Group membership/Relations to others

(continued)

Table 2 (continued)

Category	Examples
Racial or ethnic origin	Ancestors
	Language
	Ethical education
Political opinion	Political orientation
	Party affiliation
	Votes given in elections
Beliefs	Religious beliefs
	Philosophical beliefs
Trade union membership	Membership
	Interests
Health data	Diseases
	Medication
	Predispositions
	Blood test results
	X-ray images
Sexuality	Sex life
	Sexual orientation

to build models of optimal health against which individual conditions can be bench-marked. The potential benefits are considerable. It may become possible to improve therapies, prevent certain health problems, or extend the human life span by slowing the ageing process (Bruynseels et al., 2018).

In other areas, the integration of the genetic and biographic codes is only in its infancy. Nonetheless, powerful predictive models that rely mostly on biographic data are in place already. Merging them with information from the genetic code may dramatically change the nature of predictive models.

For example, when internet users rely extensively on applications marketed by Google, they leave a wealth of data with a private company. Google possesses information on a person's mobility patterns, consumer preferences and actual sales, sleeping times, holiday locations, private circumstances, dress style, musical taste, sexual orientation, sorrows, health problems, job losses, and accidents. Google also offers analytic tools for other private companies, and the power to predict individual consumer preferences hinges on integrating or 'synergising' data (Park & Skoric, 2017). When modern data collection methods involving glasses, watches, or other wearables create user profiles, biographical data will become even more fine-grained. Linking genetic information to this wealth of data would dramatically increase the predictive potential of models based on data collected by Google.

The Chinese social credit system is not meant to improve marketing for private companies; instead, it is a data collection initiative designed to discipline citizens. In 2019, the Chinese government planned to implement an overall quantitative rating for each citizen based on allegedly good behaviour (Kshetri, 2020). The rating, and similar ratings that exist already, will influence an individual's prospects in terms of credit availability or jobs. Much like Google, the various agents involved are combining data of different types from various sources. The information that the Chinese government appears to be interested in includes donations to charities, caring for parents, books read, opinions on the government published on the internet, walking a dog without a leash, smoking, and playing music too loud on the train (Kshetri, 2020). These data come from reports by private companies and state agencies, including courts, but also from the constant surveillance of citizens. It is estimated that 800,000 public cameras are in place in Beijing alone (Giesen, 2019). Much like Google's data pool, the Chinese social credit system consists of biographic data. Linking genetic data would likely allow an authoritarian government to predict and prevent deviant behaviour.

Another example concerns career platforms such as LinkedIn. On these platforms, individuals brand themselves as independent professionals, entrepreneurs, students, or employees by providing information on their educational attainment, skills, and career steps. By scraping the resulting knowledge pool and analysing the data based on appropriate theories, third parties such as employers can gain important information on career patterns. Dlouhy and Froidevaux (2021), for example, reconstructed different career trajectories and highlighted the differences between men and women. It is likely that in the future, career websites will become more important and include much more information. This is because skilled professionals will probably be obliged by market forces to build a personal brand on such websites. (For more information on career platforms as an application of human digital twins, see Sect. 4.2.) From our point of view, which we describe in this chapter, genetic characteristics or verified personality traits are clear candidates for information items that will be disclosed on LinkedIn and similar sites. In other words, individual digital twins on websites will grow, offering more opportunities for data analysis.

3.3   Prediction and Human Free Will

As shown in the previous examples, prediction is to some extent possible already, and it will become more accurate as the genetic and biographic codes are combined. Reliable prediction would represent important progress for cyber-physical systems, which, among other things, are meant to simulate processes in, e.g., industrial production, consumerism, or societal behaviour. However, the development of a reliable DT is extremely difficult since human behaviour arises from a complex interplay between individual, social, and material influences. Even if a model encompassing all these factors were available, it would still be impossible to predict human behaviour in a deterministic way, thanks to the central characteristic of human beings: their free will. Humans are often aware that data are being compiled. Free will enables them to change their behaviour, even against their own regular preferences, which are usually a part of the regular model of a human. Additionally, life is contingent and full of coincidences that cannot be captured in predictive models. These two aspects, coincidence and free will, clearly show that the (global) digital twins of humans differ from machine digital twins. In the following, two major problems with modelling the drivers of human behaviour will be discussed. The following factors both arise from the idea of free will: individual needs and the origins of human skills.

3.3.1 Basic Human Needs

Striving for autonomy is one of human beings' basic needs. Deci and Ryan (2012) justified this need based on an anthropological approach that implies the idea of free will. Humans tend to resist constraints and limitations on their actions and action planning, as Williams et al. (2019) revealed for the intention to donate blood[4] and as Bilal et al. (2021) demonstrated for the proactive behaviour of employees.[5] If individuals interpreted external influences in a stable and comparable way, it would perhaps be possible to develop a (somewhat) reliable predictive model of their behaviour. However, individuals interpret external influences (whether they are social or material) in ways that highly depend on the situational context. For example, workers may accept orders from colleague A and perceive orders from colleague B as a limitation of their autonomy—and these judgments may differ from one day to the next. Hence, human behaviour is mainly determined by a highly situational interplay between the individual and external contributions that is difficult to codify (see also Sect. 3.2).

3.3.2 Human Knowledge

There is strong agreement that knowledge can be considered an important contribution to human skills (Ericsson & Charness, 1994). Free will is the opportunity to choose freely between (available) courses of action (Ajzen, 2011). Thus, an appropriate model of a human as a component of a DT also needs to comprise the person's stock of knowledge, which is part of the biographical data. Analyses of knowledge structures that emerge in human performance and behaviour distinguish three domains of knowledge (Billettet al., 2018): (a) The first domain contains the kind of declarative knowledge that defines standards of performance and the canonical basis of a domain. This canonical knowledge comprises knowledge about terms, things, and rules that exist, and it is traditionally summarised in handbooks, curricula, and manuals. Hence, it is well codified and can be easily integrated into models of human behaviour. However, declarative knowledge does not automatically enable individuals to act appropriately, because it is formulated on a general level in order to make it applicable in different contexts and situations. (b) The second domain thus comprises procedural knowledge that describes how to use declarative knowledge. This kind of knowledge is highly enriched by individual experiences in various situations.

---

[4] The readiness to donate blood decreases if subjects feel that they are under pressure.

[5] Employees reduce their proactive behaviour if they feel restricted by managers' instructions.

When we apply knowledge practically and reflect on its effectiveness, we compile declarative knowledge into procedural knowledge. Highly experiential procedural knowledge also comprises routines and does not necessarily require consciousness to be acquired (Harteis & Billett, 2013). This knowledge is highly idiosyncratic and difficult or impossible to verbalise and thus to codify. (c) The third domain describes the physical and psychological representation of all knowledge stocks within a human. Even though methods of analysing the human body and brain have developed impressively, little is known about this representation of knowledge within the human body (Man & Damasio, 2019). Hence, it is obviously impossible to codify this representation appropriately.

3.3.3 Free Will and Predictions

To summarise, free will can be considered a crucial characteristic of humans. This raises the question of how free will can be implemented in a digital twin. As briefly explored, free will comprises a human's ability to act according to rules or to break the rules; it allows humans to resist their preferences and to postpone situational needs, but it also allows humans to act arbitrarily. Human decision-making occurs in highly situational ways. However, this does not mean that human beings act randomly: They can act according to their knowledge—whether it is explicit or tacit. The way people act results from their biographies, but it cannot be read as a pure extrapolation of prior experiences. Algorithms represent a mathematical technique for extrapolating decisions based on existing (or fictive) data. However, the application of automated, algorithm-based decision-making reveals that such decisions tend to be conservative rather than creative. Free will can be considered a basis for creative decisions, which are decisions that have never been made before. Though creative decisions may be made occasionally, they are not made randomly. Software-based decisions may also be novel if they are made using a random combination of logical rules. The assumption here is that free will is one of the crucial human qualities that cannot be represented in programs. The key argument is that wide parts of human knowledge remain or become tacit and cannot be verbalised. Additionally, individuals differ widely in their perceptions of the same situation. One explanation for this is that an individual's experience is saturated by individual situational perceptions that cannot be codified.

4   Digital Twin Knowledge as a Common Pool Resource: Two Tragedies

Realising the global digital twin could help individuals shape their lives in many ways, for example, by improving their health, prolonging their lifespans, and providing better training opportunities and jobs. Though each person has their own digital twin, the predictive and practical quality of each twin depends on the participation of numerous actors. Only when many persons are willing to provide data will artificial intelligence (AI) produce meaningful results; the global digital twin needs big data. Only when the many isolated models of digital twins are integrated will progress towards meaningful models evolve. Hence, universities and private firms should grant access to their data pools and predictive models. In economic terms, therefore, the knowledge on which the global digital twin of a human is built, much like other knowledge or information goods (Hess & Ostrom, 2003), is a 'common-pool resource'. Everybody contributes to the resource and benefits at the same time. A good example is the Living Heart Project (3DS, 2021). Companies, universities, and government agencies pool insights about factors contributing to various health problems related to the heart, and patients are willing to have their personal health data analysed. Using sophisticated models and big data, it becomes possible to simulate, for example, the factors leading to heart problems. The knowledge is a common pool, brought together by many actors, but each patient benefits individually because their own digital data can be combined with general insights, creating a well-founded individual digital twin. Each patient can be treated in appropriate ways without going through extensive screenings. Each patient contributes by consenting to the use of their own personal health data and in turn benefits from better treatment and health.

Such common pool resources lead to numerous incentive problems. They can be summarised by two symmetrical 'tragedies' (Buchanan & Yoon, 2000); these tragedies also plague the knowledge pool underlying the global digital twin.

On the one hand, the resource can be used in inappropriate ways that ultimately harm the resource base, possibly resulting in its exhaustion. For example, participating in the Living Heart Project may leak personal data. When this undermines the willingness of other firms and individual patients to contribute, the resource base may shrink or fail to grow sufficiently. Likewise, patients may be willing to learn about their own lives without being willing to pay for this benefit or without being willing to provide accurate health data. Again, the growth of the knowledge base is harmed by inappropriate use. On the other hand, the resource may be less valuable if some actors are excluded from accessing it. In this case, the resource is not used inappropriately, as in the first problem; instead, it is not used enough. For example, each private company may try to implement its own living heart project in order to make higher profits. However, then the possible benefits of learning from really big data may be lost. Similarly, excluding patients who are unable to pay from accessing the project reduces the sample size. Depending on how people are excluded, important groups such as children or people with prior health problems may be underrepresented in or completely absent from the data, and thus important information for the broad application of the knowledge base is lost.

Common pool resources are governed by different systems of rules, which range from private property to common property rules but mostly focus on a complex sharing of rights (Hess & Ostrom, 2003). Local projects such as the Living Heart Project can often be governed efficiently. It is more difficult to find rules for the knowledge pool underlying the global digital twin because numerous actors and data from various sources are involved. By focusing on two opposing tragedies or incentive problems, the economic model of a common pool resource yields important insights into appropriate rules for the global digital twin, and the nature of these rules will be summarised in the concluding section (see also Sect. 5).

5 Conclusions

Though the global digital twin of each human being is unique, it can only be fully realised by tapping into a common pool resource, namely the knowledge created from the analysis of big data concerning millions or even billions of people. Therefore, nobody is the sole proprietor of their own global digital twin. The rules that govern the knowledge pool must strike a balance between the two tragedies mentioned in the previous section. On the one hand, individual citizens should be entitled to privacy: They should be allowed and effectively able to deny access to their own personal data, and they should know which items of information are stored by whom. This is because citizens should be able to refuse to share information if private companies or state agencies are perceived to be using the knowledge pool inappropriately. Securing such personal rights, as the European GDPR does, will encourage more citizens to share data. On the other hand, there is the danger of the overly restrictive use of personal data. The global digital twin of a human being relies on combining data from different sources and large samples. By keeping data pools separate, we will lose the benefits that can be obtained from building better cross-disciplinary models.

The largest gains in terms of predictive power will result from examining large datasets and simultaneously deciphering the biographic and genetic codes. This calls for public-private partnerships in research because much of the data pool is being gathered by private companies. It also calls for more interdisciplinary research involving data and the behavioural and health sciences. At the moment, large private companies appear to be more capable and willing than universities to transcend the traditional boundaries between disciplines (Visholm et al., 2012).

5.1 Short Summary

Based on our findings in this chapter, we can summarise the main characteristics of a human digital twin as follows:

1. The global digital twin of a human contains all the data that can be stored about a person.
2. The global digital twin is constantly changing, and new data are being added from conception to death.
3. The data stored in the digital twin can be divided into two major areas: genetic data and biographical data. It does not matter whether this information was collected and stored intentionally or unintentionally; either way, it is an important part of the database.

4. It can be assumed that the stored data contain more information than can be read directly and that more information is created through the combination/joint analysis of data. Information obtained through analysis and joint consideration is also part of the digital twin.
5. There can be joint twins in which the data and analysis results from several people are stored together.
6. In addition to the global digital twin, there are sub-twins in which only the data relevant to a specific use case are stored.
7. Humans possess free will. Even though a great deal of information about human beings can be stored, analysed, and predicted, free will cannot be stored.
8. The digital twin of a person should only belong to the person whose information was used to create it.

Note: The complete global digital twin is a concept and may never exist.

Acknowledgements Sarah Pilz, Talea Hellweg, Christian Harteis, Ulrich Rückert, and Martin Schneider are members of the research programme 'Design of Flexible Work Environments—Human-Centric Use of Cyber-Physical Systems in Industry 4.0', which is supported by the North Rhine-Westphalian funding scheme 'Forschungskolleg'. We would like to thank Marc Wollny for his assistance with literature research.

References

3DS. (2021). The Living Heart Project. https://www.3ds.com/products-services/simulia/solutions/life-sciences-healthcare/the-living-heart-project/

Ajzen, I. (2011). The theory of planned behaviour: Reactions and reflections.

Alderson, J., & Johnson, W. (2016). The personalised 'digital athlete': An evolving vision for the capture, modelling and simulation, of on-field athletic performance. In *ISBS-conference proceedings archive*.

Bagaria, N., Laamarti, F., Badawi, H. F., Albraikan, A., Martinez Velazquez, R. A., & El Saddik, A. (2020). Health 4.0: Digital twins for health and well-being. In *Connected health in smart cities* (pp. 143–152). Springer.

Barricelli, B. R., Casiraghi, E., & Fogli, D. (2019). A survey on digital twin: Definitions, characteristics, applications, and design implications. *IEEE Access, 7*, 167653–167671. https://doi.org/10.1109/ACCESS.2019.2953499

Barricelli, B. R., Casiraghi, E., Gliozzo, J., Petrini, A., & Valtolina, S. (2020). Human digital twin for fitness management. *IEEE Access, 8*, 26637–26664. https://doi.org/10.1109/ACCESS.2020.2971576

Baskaran, S., Niaki, F. A., Tomaszewski, M., Gill, J. S., Chen, Y., Jia, Y., Mears, L., & Krovi, V. (2019). Digital human and robot simulation in automotive assembly using Siemens process simulate: A feasibility study. *Procedia Manufacturing, 34*, 986–994.

Berisha-Gawlowski, A., Caruso, C., & Harteis, C. (2021). The concept of a digital twin and its potential for learning organizations. *Digital transformation of learning organizations* (pp. 95–114). Cham: Springer.

Bilal, M., Chaudhry, S., Amber, H., Shahid, M., Aslam, S., & Shahzad, K. (2021). Entrepreneurial leadership and employees' proactive behaviour: Fortifying self determination theory. *Journal of Open Innovation: Technology, Market, and Complexity, 7*(3), 176.

Billett, S., Harteis, C., & Gruber, H. (2018). Developing occupational expertise through everyday work activities and interactions. *The Cambridge handbook of expertise and expert performance* (pp. 105–126).

Bruynseels, K., Santoni de Sio, F., & van den Hoven, J. (2018). Digital twins in health care: Ethical implications of an emerging engineering paradigm. *Frontiers in Genetics, 9*, 31. https://doi.org/10.3389/fgene.2018.00031

Bryndin, E. (2019a). Collaboration robots with artificial intelligence (AI) as digital doubles of person for communication in public life and space. *Budapest International Research in Exact Sciences (BirEx-Journal), 1*(4), 1–11.

Bryndin, E. (2019b). Robots with artificial intelligence and spectroscopic sight in hi-tech labor market. *International Journal of Systems Science and Applied Mathematic, 4*(3), 31–37.

Buchanan, J. M., & Yoon, Y. J. (2000). Symmetric tragedies: Commons and anticommons. *The Journal of Law and Economics, 43*(1), 1–14.

Bush, V. (1945). As we may think. The Atlantic. https://www.theatlantic.com/magazine/archive/1945/07/as-we-may-think/303881/

Chakshu, N. K., Sazonov, I., & Nithiarasu, P. (2020). Towards enabling a cardiovascular digital twin for human systemic circulation using inverse analysis. *Biomechanics and Modeling in Mechanobiology*. https://doi.org/10.1007/s10237-020-01393-6

Consortium, E. P., et al. (2012). An integrated encyclopedia of DNA elements in the human genome. *Nature, 489*(7414), 57.

Croatti, A., Gabellini, M., Montagna, S., & Ricci, A. (2020). On the integration of agents and digital twins in healthcare. *Journal of Medical Systems, 44*(9).

Dahm, R. (2008). Discovering DNA: Friedrich Miescher and the early years of nucleic acid research. *Human Genetics, 122*(6), 565–581.

Davis, C. A., Hitz, B. C., Sloan, C. A., Chan, E. T., Davidson, J. M., Gabdank, I., Hilton, J. A., Jain, K., Baymuradov, U. K., Narayanan, A. K., et al. (2018). The encyclopedia of DNA elements (ENCODE): Data portal update. *Nucleic Acids Research, 46*(D1), D794–D801.

Deci, E. L., & Ryan, R. M. (2012). Self-determination theory. In Van Lange, P. A. M., Kruglanski, A. W., & Higgins, E. T. (eds) *Handbook of theories of social psychology* (vol. 1).

Dlouhy, K., & Froidevaux, A. (2021). Evolution of STEM professionals' careers upon graduation and occupational turnoverover time. In *Presented at workshop of WK personal 2021*, Düsseldorf.

Dorrer, M. (2020). The digital twin of the business process model. *Journal of Physics: Conference Series, 1679*, 032096. IOP Publishing.

Egger, G., Liang, G., Aparicio, A., & Jones, P. A. (2004). Epigenetics in human disease and prospects for epigenetic therapy. *Nature, 429*(6990), 457–463.

El Saddik, A. (2018). Digital twins: The convergence of multimedia technologies. *IEEE MultiMedia, 25*(2), 87–92. https://doi.org/10.1109/MMUL.2018.023121167

Elgan, M. (2016). Lifelogging is dead (for now). https://www.computerworld.com/article/3048497/lifelogging-is-dead-for-now.html

Engels, G. (2020). Der digitale fußabdruck, schatten oder zwilling von maschinen und menschen. *Gruppe Interaktion Organisation Zeitschrift für Angewandte Organisationspsychologie (GIO), 51*(3), 363–370. https://doi.org/10.1007/s11612-020-00527-9

Ericsson, K. A., & Charness, N. (1994). Expert performance: Its structure and acquisition. *American Psychologist, 49*(8), 725.

Erol, T., Mendi, A. F., & Doğan, D. (2020). The digital twin revolution in healthcare. In *2020 4th international symposium on multidisciplinary studies and innovative technologies (ISMSIT)* (pp. 1–7). IEEE.

Fraga, M. F., Ballestar, E., Paz, M. F., Ropero, S., Setien, F., Ballestar, M. L., Heine-Suñer, D., Cigudosa, J. C., Urioste, M., Benitez, J., et al. (2005). Epigenetic differences arise during the lifetime of monozygotic twins. *Proceedings of the National Academy of Sciences, 102*(30), 10604–10609.

Fuller, A., Fan, Z., Day, C., & Barlow, C. (2020). Digital twin: Enabling technologies, challenges and open research. *IEEE Access, 8*, 108952–108971. https://doi.org/10.1109/ACCESS.2020.2998358

Gartner. (2020). Gartner hype cycle for emerging technologies 2020. https://www.gartner.com/ smarterwithgartner/5-trends-drive-the-gartner-hype-cycle-for-emerging-technologies-2020

GDPR. (2016). EU General Data Protection Regulation (GDPR): Regulation (EU) 2016/679.

Giesen, C. (2019). Ein ganzes land als testgelände. Süddeutsche Zeitung. https://www.sueddeutsche. de/politik/china-ein-ganzes-land-als-testgelaende-1.4664052

Glaessgen, E., & Stargel, D. (2012). The digital twin paradigm for future NASA and U.S. Air force vehicles. In *53rd AIAA/ASME/ASCE/AHS/ASC structures, structural dynamics and materials conference, 20th AIAA/ASME/AHS adaptive structures conference, 14th AIAA*. Reston, Virginia: American Institute of Aeronautics and Astronautics. https://doi.org/10.2514/6.2012-1818

Gámez Díaz, R., Yu, Q., Ding, Y., Laamarti, F., & El Saddik, A. (2020). Digital twin coaching for physical activities: A survey. *Sensors, 20*(20), 5936.

Gomerova, A., Volkov, A., Muratchaev, S., Lukmanova, O., & Afonin, I. (2021). Digital twins for students: Approaches, advantages and novelty. In *2021 IEEE conference of russian young researchers in electrical and electronic engineering (ElConRus)* (pp. 1937–1940). https://doi. org/10.1109/ElConRus51938.2021.9396360

Graessler, I., & Poehler, A. (2018). Intelligent control of an assembly station by integration of a digital twin for employees into the decentralized control system. *Procedia Manufacturing, 24*, 185–189. https://doi.org/10.1016/j.promfg.2018.06.041

Grieves, M. (2015). Digital twin: Manufacturing excellence through virtual factory replication. Whitepaper.

Hafez, W. (2020). Human digital twin: Enabling human-multi smart machines collaboration. In Y. Bi, R. Bhatia, & S. Kapoor (Eds.), *Intelligent systems and applications* (pp. 981–993). Cham: Springer International Publishing.

Hafez, W. (2020b). Human digital twins: Two-layer machine learning architecture for intelligent human-machine collaboration. In *International conference on intelligent human systems integration* (pp. 627–632). Springer.

Harteis, C., & Billett, S. (2013). Intuitive expertise: Theories and empirical evidence. *Educational Research Review, 9*, 145–157.

Heinke, A. (2021). How humans and machines interact. In S. Güldenberg, E. Ernst, & K. North (Eds.), *Managing work in the digital economy: Challenges, strategies and practices for the next decade* (pp. 21–39). Cham: Springer International Publishing.

Hess, C., & Ostrom, E. (2003). Ideas, artifacts, and facilities: Information as a common-pool resource. *Law and Contemporary Problems, 66*(1/2), 111–145.

Hintz, A., Dencik, L., & Wahl-Jorgensen, K. (2018). *Digital citizenship in a datafied society*. Wiley.

Huypens, P., Sass, S., Wu, M., Dyckhoff, D., Tschöp, M., Theis, F., Marschall, S., de Angelis, M. H., & Beckers, J. (2016). Epigenetic germline inheritance of diet-induced obesity and insulin resistance. *Nature Genetics, 48*(5), 497–499.

Jimenez, J. I., Jahankhani, H., & Kendzierskyj, S. (2020). Health care in the cyberspace: Medical cyber-physical system and digital twin challenges. In *Digital twin technologies and smart cities* (pp. 79–92). Springer.

Johnson, W. R., Mian, A., Donnelly, C. J., Lloyd, D., & Alderson, J. (2018). Predicting athlete ground reaction forces and moments from motion capture. *Medical & Biological Engineering & Computing, 56*(10), 1781–1792.

Jones, D., Snider, C., Nassehi, A., Yon, J., & Hicks, B. (2020). Characterising the digital twin: A systematic literature review. *CIRP Journal of Manufacturing Science and Technology, 29*, 36–52. https://doi.org/10.1016/j.cirpj.2020.02.002

Joseph, A., Kruger, K., & Basson, A. H. (2020). An aggregated digital twin solution for human-robot collaboration in Industry 4.0 environments. In *International workshop on service orientation in holonic and multi-agent manufacturing* (pp. 135–147). Springer.

Kemény, Z., Beregi, R., Nacsa, J., Glawar, R., & Sihn, W. (2018). Expanding production perspectives by collaborating learning factories–Perceived needs and possibilities. *Procedia Manufacturing, 23*, 111–116.

Kesti M (2021) The digital twin of an organization by utilizing reinforcing deep learning. In *Artificial neural networks and deep learning-Applications and perspective*, IntechOpen.

Kshetri, N. (2020). China's social credit system: Data, algorithms and implications. *IT Professional, 22*(2), 14–18.

Laamarti, F., Badawi, H. F., Ding, Y., Arafsha, F., Hafidh, B., & Saddik, A. E. (2020). An ISO/IEEE 11073 standardized digital twin framework for health and well-being in smart cities. *IEEE Access, 8*, 105950–105961. https://doi.org/10.1109/ACCESS.2020.2999871

Labonté, B., Suderman, M., Maussion, G., Navaro, L., Yerko, V., Mahar, I., Bureau, A., Mechawar, N., Szyf, M., Meaney, M. J., et al. (2012). Genome-wide epigenetic regulation by early-life trauma. *Archives of General Psychiatry, 69*(7), 722–731.

Lim, K. Y. H., Zheng, P., & Chen, C. H. (2020). A state-of-the-art survey of digital twin: Techniques, engineering product lifecycle management and business innovation perspectives. *Journal of Intelligent Manufacturing, 31*, 1313–1337. https://doi.org/10.1007/s10845-019-01512-w

Man, K., & Damasio, A. (2019). Homeostasis and soft robotics in the design of feeling machines. *Nature Machine Intelligence, 1*(10), 446–452.

Matusiewicz, D., Puhlalac, V., & Werner, J. A. (2018). Avatare im gesundheitswesen. https://www.youtube.com >watch<v=g7Bxm60B-kc

Microsoft Research. (2017). Mylifebits - Microsoft Research. https://www.microsoft.com/en-us/research/project/mylifebits/

Mossberger, K., Tolbert, C. J., & McNeal, R. S. (2007). *Digital citizenship: The Internet, society, and participation*. MIT Press.

NIH. (2018). Genetics vs. genomics fact sheet. https://www.genome.gov/about-genomics/fact-sheets/Genetics-vs-Genomics

NIH. (2020). Human genome project FAQ. https://www.genome.gov/human-genome-project/Completion-FAQ

Nikolakis, N., Alexopoulos, K., Xanthakis, E., & Chryssolouris, G. (2019). The digital twin implementation for linking the virtual representation of human-based production tasks to their physical counterpart in the factory-floor. *International Journal of Computer Integrated Manufacturing, 32*(1), 1–12.

Park, Y. J., & Skoric, M. (2017). Personalized ad in your Google Glass? Wearable technology, hands-off data collection, and new policy imperative. *Journal of Business Ethics, 142*(1), 71–82.

Petzoldt, C., Wilhelm, J., Hoppe, N. H., Rolfs, L., Beinke, T., & Freitag, M. (2020). Control architecture for digital twin-based human-machine interaction in a novel container unloading system. *Procedia Manufacturing, 52*, 215–220.

Popa, E. O., van Hilten, M., Oosterkamp, E., & Bogaardt, M. J. (2021). The use of digital twins in healthcare: Socio-ethical benefits and socio-ethical risks. *Life Sciences, Society and Policy, 17*(1), 1–25.

Rodríguez Aguilar, R., & Marmolejo Saucedo, J. A. (2020). Conceptual framework of digital health public emergency system: Digital twins and multiparadigm simulation.

Shengli, W. (2021). Is human digital twin possible? *Computer Methods and Programs in Biomedicine Update, 1*. https://doi.org/10.1016/j.cmpbup.2021.100014

Sun, J., Tian, Z., Fu, Y., Geng, J., & Liu, C. (2021). Digital twins in human understanding: A deep learning-based method to recognize personality traits. *International Journal of Computer Integrated Manufacturing, 34*(7–8), 860–873. https://doi.org/10.1080/0951192X.2020.1757155

Suzuki, M. M., & Bird, A. (2008). DNA methylation landscapes: Provocative insights from epigenomics. *Nature Reviews Genetics, 9*(6), 465–476.

Sweeney, L. (2000). Simple demographics often identify people uniquely. *Health (San Francisco), 671*(2000), 1–34.

Terpsma, R. J., & Hovey, C. B. (2020). *Blunt impact brain injury using cellular injury criterion*, Technical report. Sandia National Lab (SNL-NM), Albuquerque, NM (United States).

Thumfart, K. M., Jawaid, A., Bright, K., Flachsmann, M., & Mansuy, I. M. (2021) Epigenetics of childhood trauma: Long term sequelae and potential for treatment. *Neuroscience & Biobehavioral Reviews*.

Tröbinger, M., Jähne, C., Qu, Z., Elsner, J., Reindl, A., Getz, S., Goll, T., Loinger, B., Loibl, T., Kugler, C., et al. (2021). Introducing GARMI-A service robotics platform to support the elderly at home: Design philosophy, system overview and first results. *IEEE Robotics and Automation Letters, 6*(3), 5857–5864.

Truby, J., & Brown, R. (2021). Human digital thought clones: The Holy Grail of artificial intelligence for big data. *Information & Communications Technology Law, 30*(2), 140–168.

Visholm, A., Grosen, L., Norn, M. T., & Jensen, R. L. (2012). *Interdisciplinary research is key to solving society's problems*. DEA, Copenhagen Interdisciplinarity and Sustainability: Shaping Futures.

Voigt, I., Inojosa, H., Dillenseger, A., Haase, R., Akgün, K., & Ziemssen, T. (2021). Digital twins for multiple sclerosis. *Frontiers in Immunology, 12*, 1556.

Watson, J. D., & Crick, F. H. (1953). Molecular structure of nucleic acids: A structure for deoxyribose nucleic acid. *Nature, 171*(4356), 737–738.

Wetterstrand, K. A. (2020). DNA sequencing costs: Data from the NHGRI genome sequencing program (GSP). www.genome.gov/sequencingcostsdata

Williams, L. A., Sun, J., & Masser, B. (2019). Integrating self-determination theory and the theory of planned behaviour to predict intention to donate blood. *Transfusion Medicine, 29*, 59–64.

Yigitbas, E., Karakaya, K., Jovanovikj, I., & Engels, G. (2021). Enhancing human-in-the-loop adaptive systems through digital twins and VR interfaces. arXiv:2103.10804

Zibuschka, J., Ruff, C., Horch, A., & Roßnagel, H. (2020). A human digital twin as building block of open identity management for the Internet of Things. Open Identity Summit 2020.

Planning of the Digital and Networked Work

Enhancing Risk Management for Digitalisation Projects in the Context of Socio-Technical Systems

Jörn Steffen Menzefricke, Christian Koldewey, and Roman Dumitrescu

Abstract Digitalisation offers companies the potential to shape their processes and services along the entire value chain in a more efficient, flexible, and resource-saving way. However, there are many risks that companies need to consider in their digitisation efforts. These risks include not only the risks associated with technical challenges but also the risks concerning the organisation and the people working within it, such as, e.g., a lack of competencies or inappropriate processes. Even if new technologies are successfully implemented, their effective performance depends to a large extent on compatible processes, the relevant competencies of employees, and employees' acceptance. Therefore, it is important to identify the risks associated with digital transformation at an early stage so that necessary measures for risk mitigation can be planned and implemented. In this way, the appropriate conditions for successful digitalisation within the company can be established. To meet these challenges, it is necessary to move from traditional risk management to holistic socio-technical risk management. This chapter shows how the classic risk management cycle can be enhanced for digitalisation projects against a socio-technical background. In addition, methods and approaches are explained that can help to support the risk management phases of identification and analysis, assessment, and treatment in a targeted manner.

Keywords Risk management · Digitalisation · Socio-technical systems

J. S. Menzefricke (✉) · C. Koldewey · R. Dumitrescu
Heinz Nixdorf Institute, University of Paderborn, Fürstenallee 11, 33102 Paderborn, Germany
e-mail: menzefricke@hni.uni-paderborn.de

C. Koldewey
e-mail: christian.koldeway@hni.uni-paderborn.de

R. Dumitrescu
e-mail: roman.dumitrescu@hni.uni-paderborn.de

R. Dumitrescu
Fraunhofer Institute for Mechatronic Systems Design, Zukunftsmeile 1, 33102 Paderborn, Germany

1 Introduction

The rapidly increasing interconnection of physical objects and actors via the so-called Internet of Things has opened up a wide range of potential benefits for companies related to digitalisation (Kagermann, 2015). By bundling the individual potentials of digital technologies in a targeted manner, companies can continuously improve the performance of their value creation processes and their market performances (BMWi, 2015). To this end, companies need to coordinate individual digitalisation projects in a synergetic way and thus undergo a holistic digital transformation (Kofler, 2018). However, the digital transformation of a company is not a trivial undertaking. Necessary changes for the successful completion of the digital transformation must be taken into account in an equal manner for the organisation, employees, and technical infrastructure (Veile et al., 2020). The interrelationships between the organisation, human, and technology dimensions are summarised using the term 'socio-technical system' (Hirsch-Kreinsen & ten Hompel, 2017). Examples of inter-relationships include human-machine interactions, digital process alignment, and the design of organisational conditions for employees. The coordinated implementation of necessary changes involves a wide range of risks that companies must address in a structured manner when they plan and implement digital transformation initiatives (Aven & Ylönen, 2018). These risks must be identified, analysed, evaluated, and finally addressed with certain measures. Managing risks is a key challenge for companies when they plan and implement a digital transformation (Boehm & Smith, 2021). Hence, far-reaching changes and the resulting interactions between the organisation, employees, and the technologies used must be taken into account within risk management (Brocal et al., 2019).

Traditional risk management provides many tools and frameworks that are already established for specific disciplines. However, the socio-technical interactions prevailing in the digital transformation require a cross-disciplinary mindset within risk management because all areas of the company are affected by digitalisation and are equally threatened by risks. This chapter looks at the socio-technical challenges of digital transformation and discusses the implications these challenges have for risk management. Furthermore, a process model is presented that shows how socio-technical risks can be identified, evaluated, and addressed, with options for action.

The procedure presented in this chapter is explained using an example. The introduction of operational data management in production serves as an example in which the usage or operating data of the machines operated by the employees are recorded and evaluated for optimisation purposes. This allows production processes to be optimised, and it allows errors and inefficiencies to be identified and eliminated. It can thus contribute to the optimisation of processes within production. However, this can also create socio-technical risks. Based on the collected data, a digital twin of an employee can be created; this allows conclusions to be drawn about the behaviour of the employee during his activities. This includes errors in operation, downtime due to equipment modification, etc. The collection of data enables the comparison and evaluation of employees. In particular, this can lead to rejection by employees. The fear of being evaluated and compared with other employees causes anxiety for many employees. They may worry about losing their jobs or being made responsible for mistakes. The lack of acceptance of operational data management in production can therefore be cited here as an exemplary risk that can jeopardise the implementation of such a solution. Therefore, it is necessary to investigate risks, especially in the context of socio-technical risk management, to counteract risk sources and effects with suitable measures.

This chapter first describes the underlying problem of linking classical risk management with ideas from socio-technical systems and briefly presents existing state-of-the-art approaches. Subsequently, fields of action for socio-technical risk management are derived. Then, these fields of action are addressed, and an example is used to explain the three essential steps of risk management with socio-technical aspects.

2 Socio-technical Characteristics of the Digital Transformation and Socio-Technical Challenges

The following section clarifies the basic understanding of digital transformation. It then explains socio-technical perspectives that can be used to structure digitisation projects.

2.1 Digital Transformation—An Overview

Companies are attempting to bundle individual digitalisation initiatives into the holistic digitalisation of an entire company in a synergetic and consistent manner. The digital transformation has the potential to help manufacturing companies to increase their productivity and thus also their competitive strength. However, existing experience with digitalisation and automation technologies varies greatly from company to company. Small and medium-sized companies have to manage with limited resources and are forced to select individual digitalisation measures. In this way, they can improve their competitiveness in a focused manner, reduce deficiencies in their value creation and development activities, and build up new business areas (Wildemann, 2018). The necessary conditions for this digitalisation force companies to make difficult investment decisions. The determination of suitable value creation areas and the identification of realisable potentials are a major challenge in this context. In order to be able to initiate specific digitalisation measures in a targeted manner, the technological maturity of a company is often determined in order to identify the relevant fields of action for digitalisation. Three generic thrusts exist for digital transformation: the digitalisation of business models, the digitalisation of market performance, and the digitalisation of value creation processes (Lipsmeyer, 2021).

Digitalisation is causing the fundamental replacement and further development of successful business models for traditional manufacturing companies; for example, many companies are making the transition from being producers of capital goods to being manufacturing service providers or platform operators for professional services (Gausemeier et al., 2017; Drewel et al., 2021). In addition, the intelligent analysis of data will enable new types of business models.

Digitalisation and the networking of products and services have the potential to improve a company's market performance in terms of both revenue and costs (Hirsch-Kreinsen, 2020). This is often achieved by involving customers in the product development process at an early stage. In addition, a better understanding of customer behaviour can be achieved by analysing the usage and operational data. These insights can in turn be incorporated into the product development of new generations or products in order to make new products more competitive (Meyer et al., 2021).

The potentials of digitalisation are primarily related to manufacturing and value creation processes. By collecting and evaluating machine data, the manual search for cause-and-effect mechanisms can be replaced by a systematic determination of statistical correlations and patterns. The use of cyber-physical systems such as cooperating robots in manufacturing leads to the increased autonomy of production systems. Production systems are linked with each other, as well as with products and with transport technologies. A network of 'smart objects' in which the sequence of machining processes and corresponding logistical functions are organised and controlled autonomously is created (Wildemann, 2018).

The potential of digital transformation can be exploited in all areas of value creation to increase the efficiency and flexibility of companies. In order to use this potential, however, the necessary conditions must be created. These conditions do not merely involve taking existing technical interfaces into account and adapting them where necessary; instead, they concern giving equal consideration to all aspects involved in the introduction of new technology. The term socio-technical systems is established as a structure-providing framework for these aspects. The challenges associated with this process are described in the following subsection.

2.2   Socio-technical Perspectives on the Digital Transformation

Many companies hesitate to implement digital solutions. This is often due not only to technical challenges but also to the effects in terms of the organisational and human factors, which are difficult to assess (Hobscheidt et al., 2020). Even if new technologies are implemented successfully, their efficient use depends on the correct adaption of the affected processes, the enhancement of competencies on the employee side, and the acceptance of the employees. A company thus represents a so-called socio-technical system. Socio-technical system theory links technological, organisational, and human elements in a superordinate context. In the literature,

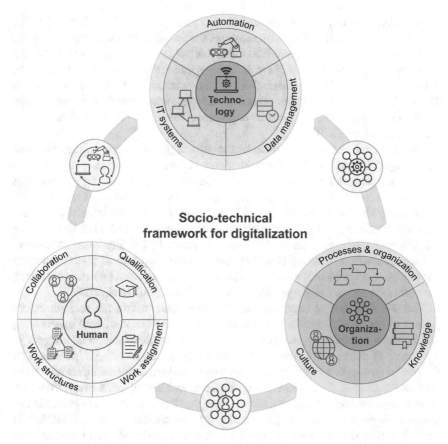

Fig. 1 Socio-technical structuring framework according to Hobscheidt et al. (2020)

there are different understandings of the components to be considered within the people, organisation, and technology dimensions. We use the socio-technical structuring framework from Hobscheidt et al. (2020) (see Fig. 1). Based on a literature analysis, this provides a generally applicable reference for the relevant components of the three socio-technical dimensions.

The technology dimension covers the automation, IT systems, and data management components. Automation can be characterised by features such as the level of automation or the use of information and communication technology for process automation. The IT systems component refers to, among other things, the IT infrastructure of a company. Features such as software support, consistent interface design, and technologies for IT security are relevant here. The data management component addresses the level and appropriate use of data (Hobscheidt et al., 2020).

The human dimension consists of the following components: qualification, work tasks, collaboration, and work structures. The qualification component deals with the education and training of employees and focuses on the competencies and skills of employees, as well as learning and qualification processes (see Sect. 4.2). The work tasks component includes, for example, the features of personal responsibility, work content, and work requirements. The collaboration component refers to the interactions between the employees themselves and the machines. In particular, the degree of interdisciplinarity and intuitive human-machine interaction are considered. The work structures component takes into account the conditions at the workplace, for example, by including ergonomic requirements or aspects of occupational safety (Hobscheidt et al., 2020).

The organisation dimension includes the culture, processes and organisation, and knowledge components. The culture component refers to the corporate culture of the organisation and addresses employee participation, sensitivity to change, and error culture. The focus of the processes and organisation component is the company's process organisation. Features such as process transparency, the integration of new business processes, and process networking are examined in this component. The handling of knowledge in the company is represented by the knowledge component in the organisation dimension. Aspects such as the existence of interdisciplinary knowledge in the company, as well as knowledge generation and knowledge transfer, are taken into account to shape this component (Hobscheidt et al., 2020).

The large number of components to be considered within the socio-technical system shows that the introduction of digital solutions is not a trivial undertaking and involves far-reaching changes. The successful implementation of projects in the course of digital transformation therefore requires far-reaching adjustments on the part of employees, organisational structures, and the technical infrastructure of companies. Only if, for example, adequate competencies, suitable work structures, suitable organisational processes, and appropriate IT environments are available can the promising implementation of digitisation projects succeed. Otherwise, employees may be overburdened with new activities, communication between business units may be faulty due to unsuitable collaboration models, or the integration of new technologies may be impeded by an unsuitable existing IT infrastructure. To ensure that necessary changes do not give rise to risks that jeopardise success, companies must become aware of these risks at an early stage and develop adequate measures for dealing with them. The following section explains how risk management works in principle and what special factors digitalisation implies in this context.

3   Risk Management in the Context of Digitalisation

The socio-technical approach is particularly relevant for risk management because it allows changes associated with digital transformation to be described systematically. In order to make this clear, traditional risk management processes are presented here; they are then expanded later to include aspects of the socio-technical approach. In the following section, the fundamentals of risk management and their theoretical correlations are explained. The challenges in the context of digital transformation are then derived.

3.1   Principles of Risk Management

The term risk management encompasses 'coordinated activities to guide and control an organization with respect to risks' (DIN31000, 2018). The ISO 31000 risk management standard establishes guidelines for dealing with risks. A risk is defined in the standard as an 'impact of uncertainty on objectives'. In this context, an impact is understood as a 'deviation from the expected', which can initially be positive or negative. Risks arise from risk causes, which can occur alone or together with other factors (DIN31000, 2018).

The ISO standard establishes a risk management framework, risk management principles, and a risk management reference process. The efficient, effective, and consistent management of risk requires the implementation or adaptation of these three components in the organisation. The principles described in the standard are intended to guide an organisation in effective risk management. The focus is on creating and protecting value by integrating risk management into all the activities of an organisation. A risk management framework with the components of leadership, integration, design, implementation, evaluation, and improvement supports this process (DIN31000, 2018).

The application of risk management is described using the reference process shown in Fig. 2. In practice, this process is not executed sequentially (as shown); instead, it is executed iteratively. The first step of the reference process is to define the application context and environment to be considered. The second step is the risk evaluation, which includes identification, analysis, and assessment. First, risks are identified. Then, risks are described and characterised based on causes and effects. Within the risk assessment, risks are evaluated on the basis of their probability of occurrence and the severity of their impact in order to identify particularly critical risks. The third step is risk treatment. This comprises the selection of measures to reduce the probabilities and the effects of risk causes (DIN31000, 2018).

These three steps are supported by secondary activities to ensure the success of the risk management process. For example, regular communication and consultation provide opportunities to involve relevant stakeholders and thereby capture sufficient information, perspectives, and expertise at each step of the risk management process. Monitoring and reviewing ensure the quality and effectiveness of the risk management process. Recording and reporting enable the communication of the results across the organisation (DIN31000, 2018).

Specific Risk Management Approaches in the Context of Digitalisation

Building on and complementing the guidelines of this risk management standard, numerous other approaches exist that often address a specific focus of risk management. Chakrabortty et al. (2019) developed a framework for assessing project risks with a focus on time and resource uncertainties. Ebert (2018) focused on risk analysis in production processes to identify technical risks such as quality defects and malfunctions. Kobi (2012) described various aspects of personnel risk management, including bottleneck risks (lack of personnel with key competencies), exit risks (key

Fig. 2 Reference process
for risk management (ISO
31000:2018)

competency personnel at risk of absconding), adjustment risks (improperly qualified personnel), and motivation risks (retaining work performance). Deng et al. (2019) dealt with the propagation effects of risks along a supply chain. Birkel et al. (2019) illustrated the need to consider the introduction of digital technologies with a holistic risk management approach, primarily due to the increasing interactions of risks from different disciplines. Hopfener and Bie (2018) considered risk management against the backdrop of digitalisation and found that risk management would play a new role in the future because of digitalisation. The authors expected that the importance of the advisory function of risk management would increase. The knowledge gained from risk management can in turn be used to continuously review and adjust the corporate strategy. In addition to this new role, a change in risk management methods is expected. This also allows risk management to be more involved in strategic issues and to contribute to value creation (Hopfener & Bier, 2018).

Furthermore, a systematic literature review has shown that there are currently no approaches that address all phases of risk management while considering socio-technical aspects. In this review, a total of 1734 papers were analysed. Through structured assessment steps, 32 relevant articles were identified and analysed in more detail.

Contributions that study the analysis phase ascribe increasing importance to the socio-technical systems approach with human, technology, and organisation dimensions (Palvia et al., 2001; Jean-Jules & Vicente, 2020; Organ & Stapleton, 2016). Therefore, in many of these contributions, there is a consideration of all socio-technical dimensions. Due to the increasing expansion of value chains to value creation networks, risk analysis in particular has become very important for new technologies in supply chain management (Palvia et al., 2001; Deng et al., 2019; Dey et al., 2013; Keizera et al., 2002).

Most approaches that deal with risk assessment focus on the technology and organisation dimensions (Hirman et al., 2019; Kara et al., 2020). The digitalisation of supply chain management is often analysed using risk assessment methods. Obviously, in this case, the focus is on the organisation dimension.

Within risk treatment, risk strategies are often used to deal with risk (DIN31000, 2018; Ivanov et al., 2019; Tupa et al., 2017). Today, these risk strategies are often used for special corporate areas such as IT or sustainability (DIN31000, 2018; Ivanov et al., 2019; Tupa et al., 2017). Approaches for risk treatment in these areas often consider all socio-technical dimensions, albeit at a very generic level (DIN31000, 2018; Ivanov et al., 2019; Tupa et al., 2017). However, a comprehensive operationalisation of risk management strategies is necessary. Against this background, there is still a lack of suitable concepts (Brocal et al., 2019).

For more detailed information, see Menzefricke et al. (2021).

3.2 Fields of Action for Socio-Technical Risk Management

Digital transformation poses new challenges for risk management, particularly in the context of socio-technical aspects. Through the implementation of digitalisation projects, changes occur in the human, technology, and organisation dimensions; these changes can cause the emergence of risks. For example, the implementation of operational data management in production can bring about a change in work tasks. This results in new competence requirements for employees. At the same time, the processes also change, which can alter the structure of the organisation by dissolving departmental boundaries. Both changes can have a negative impact on employee acceptance, for example. Risks also arise at the technological level; for example, the failure of an automated guided vehicle (AGV) system can occur due to a production error, which increases the negative impact on acceptance. The example shows the strong interconnectedness of risk causes between the individual socio-technical dimensions, which intensifies the threat of the failure of the digitalisation project. However, interconnectedness represents only one of many reasons that companies have difficulties with digital transformation (Staufen, 2019; Boehm & Smith, 2021). For example, a study by Boehm and Smith (2021) confirmed that risk management has so far failed to keep pace with digital transformation. In particular, the lack of understanding of the complex interactions between the different terms in risk management in combination with the socio-technical dimensions is a complexity driver and the reason that classical risk management needs to be adapted for digital transformation activities. In the following, the interrelationships of the risk management terms are explained against the background of the socio-technical dimensions.

The risk management framework illustrates the relationships of relevant risk management terms. The model should be run starting with the digitalisation project and then in a top-down fashion along the lines of analysis, assessment, and treatment. The starting point and central element for organising the risk management terms is the digital transformation project to be implemented (left side of Fig. 3). The project enables the specification of risks by describing the causes, risks, and effects applicable to the use case. In accordance with the risk chain, the causes give rise to a

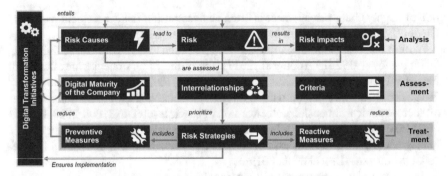

Fig. 3 Risk management framework according to (Menzefricke et al., 2021)

risk, which in turn gives rise to an impact. The recorded risks are then subjected to a socio-technical assessment (second line). The assessment is made up of the digital maturity of the company and the interactions between the risk types and identified socio-technical assessment criteria (second line). The results of the assessment are prioritised risks and critical components of the socio-technical system, which can be addressed with the help of suitable risk strategies. For this purpose, both preventive and reactive measures are used to ensure a low-risk implementation of the use case (third line). The terms and dependencies of the risk management framework are illustrated in Fig. 3.

In the risk chain described above, socio-technical interactions, which are represented by the terms causes, risk, and effects, play an essential role. The causes of a technical risk can arise in both the human dimension and the organisational dimension. Furthermore, risk effects simultaneously represent new causes for other risks. These interactions are illustrated in Fig. 4 by using a specific risk as an example to concretise the first line of the model. In the case of the implementation of operat-

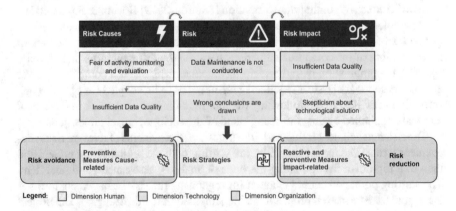

Fig. 4 Principles of risk management (illustration by the authors)

ing data management in production, for example, the necessary data maintenance is not conducted due to the fear of being monitored and evaluated. As a result, the data quality is insufficient. If this dataset is subsequently used to draw organisational conclusions, these conclusions may be incorrect. The resulting impact in turn leads to increasing scepticism about the technological performance of the systems used. This example illustrates the complex interactions that need to be taken into account in risk management in the context of digital transformation.

In addition, it is important to distinguish between preventive and reactive measures when developing measures. Preventive measures address risk causes to reduce the probability of occurrence or risk effects to keep the potential damage low when the risk occurs. Reactive measures can be implemented in the short term to minimise the resulting damage after the occurrence of a risk and to mitigate propagation effects. Against this background, the following fields of action for socio-technical risk management emerge.

Risk Identification and Analysis

The biggest problem for companies performing risk management in the context of digital transformation is the understanding of risks and their causes and effects. Companies are not aware of the risks that can arise and how these risks can spread throughout a company. While companies can often name individual risks, there are difficulties in placing these risks in an overall context (Boehm & Smith, 2021; Staufen, 2019). However, identifying interactions between causes, risks, and effects requires an overarching perspective and the ability to put these interactions into a holistic context. Therefore, a collection of generic risks (so-called risk types) and their interactions is needed; companies can use this collection of risks to identify and specify the risk types that are relevant to them (Schnasse et al., 2021). In this context, information acquisition is the most difficult phase in the entire risk management process, but at the same time it is critical to the subsequent phases (Romeike, 2018; Ellebracht et al., 2011).

Risk Assessment

Furthermore, it is important to consider risks with their interrelationships, as they are linked by complex cause-effect chains and non-deterministic dependencies. Hence, each risk is based on one or more causes and leads to various effects, which together form a chain (Romeike, 2018; Klein & Gleißner, 2017). This also applies to the interdependencies between the dimensions from the socio-technical system explained above (Baxter & Sommerville, 2011; Ulich, 2013; Anderson & Felici, 2012). Moreover, the identified risk dependencies can be used in risk assessment to gain a better understanding of the likelihood of risks occurring. In addition, further research activities can reveal whether the extent of damage from a socio-technical risk is partly due to its interactions (Schnasse et al., 2021). The degree of digital maturity can also provide information about the probability of occurrence of or the level of damage caused by a risk. By assessing the degree of digital maturity, companies can determine how prepared they are in terms of the individual socio-technical dimensions and where there is still a need for action. Various criteria are used for the evaluation to determine the degree of digital maturity of companies from a socio-technical point of view. The assessment of risks should therefore also take the degree of digital maturity into account (Menzefricke et al., 2021). Against this backdrop, the holistic identification of the socio-technical risks that arise, considering their interrelationships, is important to ensuring a sustainable digital transformation.

Risk Treatment

Since many companies are uncertain about how to deal with the risks that arise, specific risk strategies and measures for avoiding or reducing the risks must be derived on the basis of the interactions between the risks (Boehm & Smith, 2021). In this context, it is important to consider both the probability of occurrence and the potential impact if a risk materialises when choosing an appropriate course of action. Two principles of risk treatment can be identified. The principle of risk mitigation aims to minimise the impact of a risk. The reduction of the probability of occurrence is addressed by the principle of risk avoidance. In addition to these principles, there are risk acceptance, risk prevention, and risk transfer principles (Klein, 2011). In the case of risk acceptance, no actions are taken. In the case of risk prevention, reserves (e.g., financial resources) are built up for the occurrence of a risk. The transfer of a risk involves the transfer of the risk's effects to third parties, e.g., taking out insurance against a possible loss or liability clauses when working with partners (Gausemeier et al., 2017; Klein, 2011).

4 Risk Management in the Context of Socio-Technical Aspects for the Implementation of Operating Data Collection in Production

The following section shows how socio-technical risk management can be performed using the following process steps: identification and analysis, assessment, and treatment. For this purpose, the following exemplary use case is used. The collection of operating data in production can provide insights into plant usage, process errors, and fault data in the case of plant downtime. However, information about the behaviour of employees and the quality of each employee's work is also collected. This means that employees can be evaluated transparently based on the data collected and compared with one another. As a side effect, this creates a digital twin of the employee in the context of their work in production. This can lead to employees fearing that their mistakes and their behaviour at work will become transparent and thus give rise to fears of losing their jobs. This project is therefore associated with risks and is used as an example to demonstrate the methodological steps of socio-technical risk management. To this end, the basic contexts of socio-technical risk management are first presented. Then, in the following subsections, the phases of risk management (identification and analysis, assessment, and treatment) are presented from a socio-technical point of view.

4.1 Identification and Analysis with Socio-Technical Aspects

The following procedure is based on the procedure of Menzefricke et al. (2021) and is applied to the use case involving operating data management in production that is described at the beginning of this section.

As part of the risk identification process, all risks associated with the selected use case are derived, documented, and specified. A threat canvas according to Schnasse et al. (2021) is used to derive the socio-technical threats.

In the first step of risk identification (first phase of risk management), the developed threat canvas is used. For this purpose, the threats are derived along the respective components of the socio-technical dimensions. Threats are negative states or events that can result from changes. The main changes that are necessary for the implementation of the use case are recorded. The changes form the basis of the focused derivation of socio-technical risks. In the third area of the canvas, the summary and prioritisation of the derived threats is carried out. After the entire threat canvas is complete, the result is the identified threats for the respective socio-technical components under consideration (Schnasse et al., 2021). An example of a completed threat canvas can be seen in Fig. 5.

Next, causes, risks, and impacts are derived from the identified threats using a bow-tie analysis. In the bow-tie analysis, the previously developed results of the threat canvas are transformed into the causes, risks, and effects components. Guiding

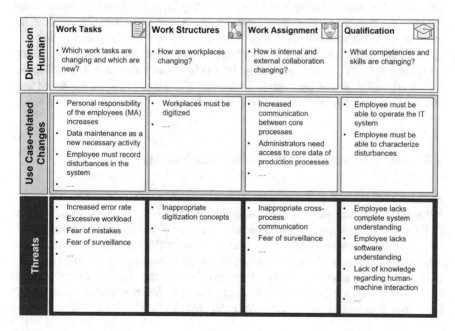

Fig. 5 Threat canvas for the human dimension according to Menzefricke et al. (2021)

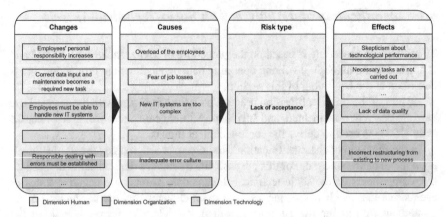

Fig. 6 Bow-tie analysis of the risk 'lack of acceptance' according to Menzefricke et al. (2021)

questions for the identification of causes include the following: Which conditions can be responsible for the occurrence of a risk? Effects can be determined with the help of the following question: What are the effects on the components of the socio-technical system? These questions are applied within a bow-tie analysis in Fig. 6.

Subsequently, the identification of interactions between the risks takes place. The procedure is based on an approach by Schnasse et al. (2021). Generic risk types were first identified on the basis of a literature analysis. Then, prevailing interactions between the risk types were determined by conducting an influence analysis. The influence analysis enables statements about how strongly a risk type, and thus the specific risk under consideration, is influenced by the other risk types. The rating scale used for this purpose comprises the values 0 (no influence), 1 (conditional influence), and 2 (strong influence). The results of the first phase are identified and analysed risks with causes, effects, and interactions.

4.2 Risk Assessment with Socio-Technical Aspects

In accordance with the risk management process, risk identification and analysis are followed by risk assessment. The aim is the identification of critical risks. To this end, the digital maturity level of a company is determined against the background of the use case implementation. This enables the identification of weaknesses in the socio-technical system. The identified risks are then assessed on the basis of their probability of occurrence and impact severity. Finally, the digital maturity level of the company and the assessed risks are compared to identify critical areas of the socio-technical system of the company under consideration.

Determining the company-specific digital maturity level

The Quick Check Industrie 4.0 of the INLUMIA research project was used as a basis for determining the maturity level. Fifty-nine criteria were adapted to the framework conditions of socio-technical risk management. A consolidation with state-of-the-art models was also carried out. This resulted in a maturity test with 96 questions covering the individual human, technology, and organisation dimensions and their components. For example, within the technology dimension, the intralogistics criterion was evaluated with the following question: 'To what extent do internal transport processes run automatically or autonomously?' The possible answers are defined using four levels and provide information about the digital maturity level. In the case of this question, the levels are manual, partially automated, automated, and autonomous. First, the company checks if the respective criterion is relevant for the use case under consideration. If it is, the company representatives answer the question and thus assess the level of digital maturity of the component. Finally, a network diagram can be created that visualises the maturity level of the company for the various components and reflects the results of the first phase (see Fig. 7).

Assessment of Use Case-Specific Risks

As a result of research on risk assessment characteristics and key data collection, the amount of damage and the probability of occurrence have emerged as established assessment dimensions.

For the assessment of the level of damage, criteria for the respective components of the dimensions were determined and consolidated in the form of an assessment sheet. For example, the impact of a risk in the 'IT systems' component is determined by the following criteria: deterioration in the system stability, the occurrence of unauthorised system access, and an increasing number of data losses/errors. The assessment includes, for example, the following question: 'How does the risk affect the system stability when it occurs?' The rating scale includes four levels. It provides statements about the damage that the company would suffer if the risk occurs within a component. At the first level of damage, only individual process steps are affected and their efficiency is restricted. In this case, the process result is not affected. The second level covers emerging damage that affects complete processes and requires intervention. The third level addresses effects that equally affect various processes and lead to a significant impairment of production activities. Level four characterises the greatest damage. In this case, the risk shuts down entire business activities and great effort is required to repair the damage.

The probability of occurrence is made up of two components: an individual assessment by company representatives of the probability of occurrence, and an influence analysis of the interactions of the associated risk types. These two components are discussed below. The company representatives individually assess the probability of occurrence of the risks with the aid of four statements, which are evaluated using a Likert scale. Thus, the statement 'There are already initial indications/risk causes for the occurrence of the risk' can be answered with 'disagree', 'tend to disagree', 'tend to agree', or 'agree'. Furthermore, the passive sum of the influence analysis is

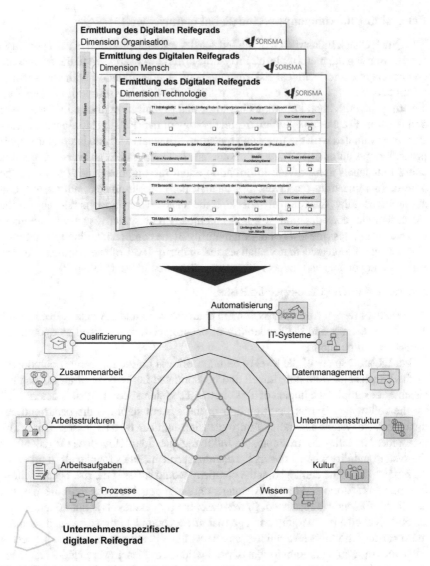

Fig. 7 Determining the degree of digital maturity (Menzefricke et al., 2021)

included in the evaluation. If a risk type is strongly influenced (high passive sum), a higher probability of occurrence is assumed (Schnasse et al., 2021). To determine the passive sum, an influence analysis of the risk types was carried out and an assessment was made of how strongly one risk type influences the others (Schnasse et al., 2021). The results of the assessment are presented in the form of a portfolio, with the probability of occurrence on the x-axis and the amount of damage on the y-axis. In addition, different threat levels within the socio-technical components of each risk can be derived from this representation.

Identification of Critical Components for Socio-Technical Risks

The assessed risks are then compared with the determined level of digital maturity of the company (see Fig. 8). For this purpose, the threat level of the risk for each component is transferred to the maturity level network diagram, creating a risk profile. By comparing this with the maturity level profile, companies can identify particularly critical components. In these components, the risk has a strong impact: It has either a high level of damage or a high probability of occurrence. At the same time, these critical components tend to have a lower level of digital maturity. The identified critical components serve as a reference object in the following phase for identifying options for action.

4.3 Risk Treatment with Socio-Technical Aspects

The object of risk control is to identify suitable options for action to reduce and limit specific risks in concrete components of the socio-technical system. Risk management has three steps: (1) the identification of the critical components of the socio-technical system, (2) the identification of options for action, and (3) the consolidation of the options for action into strategies. These three steps are described below.

Characterisation of Critical Components of the Socio-Technical System

The first step is to document the findings obtained in the previous phases for each risk. For this purpose, each risk is summarised in a risk profile. An example of a risk profile for the risk 'inappropriate data administration by employees' is shown in Fig. 9.

In total, the fact sheet contains five essential details. Each risk is described in the context under consideration (1). In addition, previously identified risk causes (e.g., the fear of monitoring and evaluation) and effects (e.g., insufficient lessons to be learned from mistakes) are assigned to the risk. The central aspect of the fact sheet is the network diagram, as well as the risk assessment. The network diagram contains the digital maturity level, as well as the determined risk profile. The assessment of the level of damage and the probability of occurrence indicates whether, within the components, the probability of occurrence or the level of damage is the more decisive dimension for a high threat level. Based on the critical components, causes, and effects, as well as the tendency of the risk assessment, options for action can be determined in the next step.

Determining Socio-Technical Options for Action

The first step in the treatment of risks with socio-technical aspects is to establish an overarching framework for action that is consistent with the digitalisation strategy. According to Menzefricke et al. (2021), socio-technical design options serve this

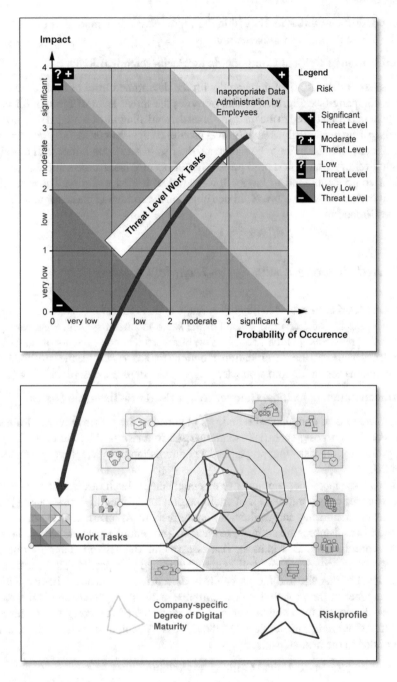

Fig. 8 Transfer of risk assessment into a threat portfolio

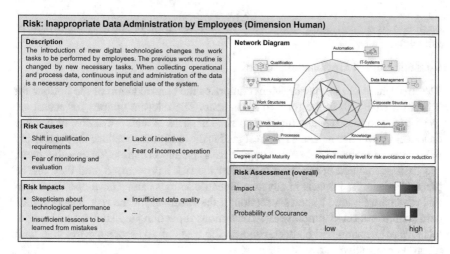

Fig. 9 Risk documentation

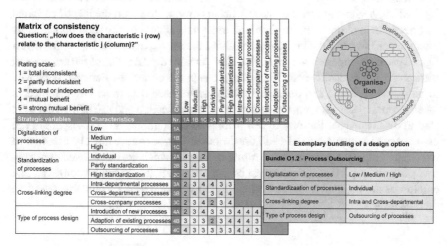

Fig. 10 Matrix of consistency

purpose. These are generic options for how a component of the socio-technical system can be designed. First, strategic variables and characteristics are identified that generically characterise the scope for action within a component. Then, consistent bundles of characteristics are identified using a consistency analysis and combined to create design options (see Fig. 10).

Consistency is then ensured by comparing the use case to be implemented with the design options. This involves checking if the characteristics of the strategic variables have already been determined by the use case and can therefore no longer be changed. This comparison makes it possible to exclude design options that are excluded by corporate or digitalisation strategies. The determination of measures can then be initiated.

Consolidation of Options for Action

The final phase of risk management comprises the consolidation of the options for action for each risk in the form of a roadmap. The relevant options for action are checked for consistency and then categorised in terms of their preventive or reactive character. Preventive measures are designed in the long term to reduce the probability of occurrence or the severity of the impact. Reactive action options, on the other hand, can be implemented in the short term after the risk has occurred, and they can be planned to address impacts in advance. Specific actions can be developed based on this roadmap. Figure 11 shows a section of the roadmap for the risk 'inappropriate data administration by employees'. The options for action are listed for each socio-technical dimension and characterised on the basis of an associated timeline with regard to preventive or reactive properties.

In this example, employee training should be initiated by management on an ongoing basis, regardless of whether the risk occurs or not. However, the identification of incorrect data and the cause of these incorrect data should be initiated immediately after the risk has occurred in order to minimise the impact of the risk. Since the investigated use case, 'operational data collection', creates digital twins of the employees, measures for the sensitive handling of these data are necessary. Personal data need to be stored securely, and unauthorised access to these data must be prevented as much as possible. It also must be ensured that the collection and use of data does not lead to disadvantages for employees. This is to be communicated transparently to employees and documented. In addition, a suitable error culture must be established to counter employee concerns. A trusting culture in which errors are used constructively to gain important insights must be established. It should not be conveyed that employees are being monitored by their digital twin and held accountable based on their behaviour.

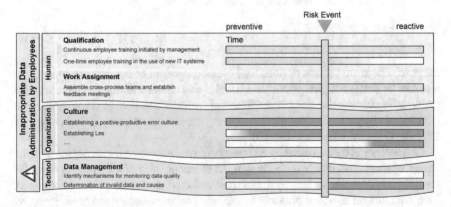

Fig. 11 Roadmap for socio-technical shaping options

5 Discussion

Digital transformation offers a wide range of potential benefits for corporate value creation. However, exploiting this potential requires far-reaching changes on the part of employees, within the organisation, and in the technologies used by companies. Companies must be able to identify the resulting risks at an early stage, assess them, and deal with them appropriately. Socio-technical interactions make it difficult for companies to apply classic risk management approaches. For this purpose, a structuring framework is presented to illustrate the relevant risk management terms and their interrelationships in a socio-technical context. Furthermore, methodological steps were developed that enable the identification, evaluation, and determination of options for action. Within the identification, changes are determined and causes, risks, and effects are derived. In the assessment, the digital maturity level of a company is considered in order to identify the endangered components of the socio-technical system in combination with the level of damage and the probability of occurrence. A set of strategic variables and characteristics was developed to identify courses of action for shaping the components at risk. Finally, a consistency assessment was used to identify courses of action that can be used to minimise or limit risks.

Nevertheless, the transferability of the method must be examined and, if necessary, it must be adapted, particularly against the background of the digitalisation of market services. To this end, the procedure should be evaluated on the basis of further market offering-related projects. In addition, risk monitoring should be integrated into the procedure as a detailed phase. Future research activities can be used to investigate what suitable monitoring mechanisms might look like and which risk indicators represent suitable metrics.

Acknowledgements Jörn Steffen Menzefricke and Roman Dumitrescu are members of the research programme 'Design of Flexible Work Environments—Human-Centric Use of Cyber-Physical Systems in Industry 4.0', which is supported by the North Rhine-Westphalian funding scheme 'Forschungskolleg'.

References

Anderson, S. D., & Felici, M. (2012). *Emerging technological risk: Underpinning the risk of technology innovation, softcover (reprint)*. London: Springer.
Aven, T., & Ylönen, M. (2018). A risk interpretation of sociotechnical safety perspectives. *Reliability Engineering & System Safety, 175*, 13–18. https://doi.org/10.1016/j.ress.2018.03.004
Baxter, G., & Sommerville, I. (2011). Socio-technical systems: From design methods to systems engineering. *Interacting with Computers, 23*(1), 4–17. https://doi.org/10.1016/j.intcom.2010.07.003
Birkel, H., Veile, J., Müller, J., Hartmann, E., & Voigt, K. I. (2019). Development of a risk framework for industry 4.0 in the context of sustainability for established manufacturers. *Sustainability, 11*(2). https://doi.org/10.3390/su11020384

BMWi. (2015). Industrie 4.0 und Digitale Wirtschaft: Impulse für Wachstum, Beschäftigung und Innovation.

Boehm, J., & Smith, J. (2021). Derisking digital and analytics transformations: While the benefits of digitization and advanced analytics are well documented, the risk challenges often remain hidden. McKinsey & Company.

Brocal, F., González, C., Komljenovic, D., Katina, P. F., & Sebastián, M. A. (2019). Emerging risk management in industry 4.0: An approach to improve organizational and human performance in the complex systems. Complexity (pp. 1–13). https://doi.org/10.1155/2019/2089763

Chakrabortty, R. K., Abbasi, A., & Ryan, M. J. (2019). A risk assessment framework for scheduling projects with resource and duration uncertainties. *IEEE Transaction Engineering Management*, 1–15. https://doi.org/10.1109/TEM.2019.2943161

Deng, X., Yang, X., Zhang, Y., Li, Y., & Lu, Z. (2019). Risk propagation mechanisms and risk management strategies for a sustainable perishable products supply chain. *Computers & Industrial Engineering, 135*, 1175–1187. https://doi.org/10.1016/j.cie.2019.01.014

Dey, P. K., Clegg, B., & Cheffi, W. (2013). Risk management in enterprise resource planning implementation: A new risk assessment framework. *Production Planning & Control, 24*(1), 1–14. https://doi.org/10.1080/09537287.2011.597038

DIN31000. (2018). *Risikomanagement - Leitlinien. 31000*. Berlin: Beuth Verlag GmbH.

Drewel, M., Özcan, L., Koldewey, C., & Gausemeier, J. (2021). Pattern-based development of digital platforms. *Creativity and Innovation Management, 30*(2), 412–430. https://doi.org/10.1111/caim.12415

Ebert, B. (2018). *Prozessoptimierung bei Industrie 4.0 durch Risikoanalysen: Gefährdungen erkennen und minimieren*. Berlin, Heidelberg: Springer.

Ellebracht, H., Lenz, G., & Osterhold, G. (2011). Systemische Organisations- und Unternehmensberatung: Praxishandbuch für Berater und Führungskräfte. *Gabler, Wiesbaden,*. https://doi.org/10.1007/978-3-8349-6920-0

Gausemeier, J. et al. (2017). Mit Industrie 4.0 zum Unternehmenserfolg: Integrative Planung von Geschäftsmodellen und Wertschöpfungssystemen.

Hirman, M., Benesova, A., Steiner, F., & Tupa, J. (2019). Project management during the Industry 4.0 implementation with risk factor analysis. *Procedia Manufacturing, 38,* 1181–1188. https://doi.org/10.1016/j.promfg.2020.01.208

Hirsch-Kreinsen, H. (2020). *Digitale Transformation von Arbeit: Entwicklungstrends und Gestaltungsansätze. Moderne Produktion*. Stuttgart: W. Kohlhammer Verlag.

Hirsch-Kreinsen, H., & ten Hompel, M. (2017). Digitalisierung industrieller Arbeit: Entwicklungsperspektiven und Gestaltungsansätze. In Vogel-Heuser B, Bauernhansl T, ten Hompel M (eds) *Handbuch Industrie 4.0* (pp. 357–376). Berlin and Heidelberg: Springer Vieweg.

Hobscheidt, D., Kühn, A., & Dumitrescu, R. (2020). Development of risk-optimized implementation paths for Industry 4.0 based on socio-technical pattern. *CIRP Design, 91*, 832–837. https://doi.org/10.1016/j.procir.2020.02.242

Hopfener, A., & Bier, S. (2018). Risikomanagement im Zeitalter der Digitalisierung: Rolle und Herausforderungen. *Risiko-Manager Fachzeitschrift für Risiko-Experten, 9*, 10–16.

Ivanov, D., Dolgui, A., & Sokolov, B. (2019). The impact of digital technology and Industry 4.0 on the ripple effect and supply chain risk analytics. *International Journal of Production Research, 57*(3), 829–846. https://doi.org/10.1080/00207543.2018.1488086

Jean-Jules, J., & Vicente, R. (2020). Rethinking the implementation of enterprise risk management (ERM) as a socio-technical challenge. *Journal of Risk Research, 24*(2), 247–266. https://doi.org/10.1080/13669877.2020.1750462

Kagermann, H. (2015). Change through digitization—Value creation in the age of Industry 4.0. In *Management of permanent change* (pp. 23–45). Wiesbaden: Springer Gabler. https://doi.org/10.1007/978-3-658-05014-6_2

Kara, M. E., Fırat, S. Ü. O., & Ghadge, A. (2020). A data mining-based framework for supply chain risk management. *Computers & Industrial Engineering, 139*. https://doi.org/10.1016/j.cie.2018.12.017

Keizera, J. A., Halman, J. I., & Song, M. (2002). From experience: Applying the risk diagnosing methodology. *Journal of Product Innovation Management, 19*(3), 213–232. https://doi.org/10. 1111/1540-5885.1930213

Klein, A. (2011). *Risikomanagement und Risiko-Controlling: Moderne Instrumente.* Haufe Fachpraxis, Haufe-Lexware, München: Grundlagen und Lösungen.

Klein, A., & Gleißner, W. (2017). *Risikomanagement und Controlling: Chancen und Risiken erfassen, bewerten und in die Entscheidungsfindung integrieren.* Haufe Lexware, München: Haufe Fachpraxis.

Kobi, J. M. (2012). *Personalrisikomanagement: Strategien zur Steigerung des People Value.* Wiesbaden: Springer Gabler.

Kofler, T. (2018). Digitale Transformation in Unternehmen: Einflusskräfte und organisatorische Rahmenbedingungen. Zentrum Digitalisierung Bayer.

Lipsmeyer, A. (2021). Systematik zur Entwicklung von Digitalisierungsstrategien für Industrieunternehmen. Dissertation, University of Paderborn, Paderborn.

Menzefricke, J. S., Gabriel, S., Gundlach, T., Hobscheidt, D., Kürpick, C., Schnasse, F., Scholtysik, M., Seif, H., Koldewey, C., & Dumitrescu, R. (2021). Soziotechnisches Risikomanagement als Erfolgsfaktor für die Digitale Transformation. In D. R. Schallmo (Ed.), *Digitalisierung: Strategie, Transformation und Implementierung.* Berlin Heidelberg: Springer.

Menzefricke, J. S., Wiederkehr, I., Koldewey, C., & Dumitrescu, R. (2021). Maturity-based development of strategic thrusts for socio-technical risks. *Procedia CIRP, 104*, 241–246. https://doi. org/10.1016/j.procir.2021.11.041

Menzefricke, J. S., Wiederkehr, I., Koldewey, C., & Dumitrescu, R. (2021). Socio-technical risk management in the age of digital transformation - Identification and analysis of existing approaches. *Procedia CIRP, 100*, 708–713. https://doi.org/10.1016/j.procir.2021.05.094

Meyer, M., Panzner, M., Koldewey, C., & Dumitrescu, R. (2021). Towards identifying data analytics use cases in product planning. *Procedia CIRP, 104*, 1179–1184. https://doi.org/10.1016/j.procir. 2021.11.198

Organ, J., & Stapleton, L. (2016). Technologist engagement with risk management practices during systems development? Approaches, effectiveness and challenges. *AI & Society, 31*(3), 347–359. https://doi.org/10.1007/s00146-015-0597-4

Palvia, S. C., Sharma, R. S., & Conrath, D. W. (2001). A socio-technical framework for quality assessment of computer information systems. *Industrial Management & Data Systems, 101*(5), 237–251. https://doi.org/10.1108/02635570110394635

Romeike, F. (2018). Risikomanagement. *SpringerLink Bücher.* Wiesbaden: Springer Gabler. https:// doi.org/10.1007/978-3-658-13952-0

Schnasse, F., Menzefricke, J. S., & Dumitrescu, R. (2021). Identification of socio-technical risks and their correlations in the context of digital transformation for the manufacturing sector. In *Proceedings of the IEEE 16th international conference on industrial electronics and applications (ICIEA).* IEEE. https://doi.org/10.1109/iciea52957.2021.9436799

Staufen, A. G. (2019). Industrie 4.0 Index: Deutscher Industrie 4.0 Index 2019. Köngen.

Tupa, J., Simota, J., & Steiner, F. (2017). Aspects of risk management implementation for Industry 4.0. *Procedia Manufacturing, 11*, 1223–1230. https://doi.org/10.1016/j.promfg.2017.07.248

Ulich, E. (2013). Arbeitssysteme als soziotechnische systeme-eine erinnerung. *Psychologie des Alltagshandelns, 6*(1), 4–12.

Veile, J. W., Kiel, D., Müller, J. M., & Voigt, K. I. (2020). Lessons learned from Industry 4.0 implementation in the German manufacturing industry. *Journal of Manufacturing Technology Management, 31*(5), 977–997. https://doi.org/10.1108/JMTM-08-2018-0270

Wildemann, H. (2018). Produktivität durch Industrie 4.0. TWC Transfer-Centrum.

Justice and Fairness Perceptions in Automated Decision-Making—Current Findings and Design Implications

Paul Hellwig and Günter W. Maier

Abstract Artificial intelligence in decision-making is a topic of increasing importance in research and society today. The concept of digital twins of humans (digital representation of humans), is also related to automated decision-making, as these digital twins could be used for decision-making. The increasing use of automated decision-making raises the question how employees perceive this form of decision-making. Therefore, research on justice and fairness perceptions is introduced in this chapter. The central part of this chapter is a literature review on justice and fairness perceptions in automated decision-making. The 43 included studies highlight the comparison of automated decision-making and human decision-making, as well as design factors and individual characteristics that influence justice and fairness perceptions. The results regarding justice and fairness perceptions are mixed. However, both decision-making authorities also offer unique strengths in terms of perceptions of justice: Automated decision-making is seen as less biased than human decision-making, while human decision-making is seen as open to peoples' views and is associated with respectful treatment. The chapter concludes with a discussion of open questions and implications for the design of automated decision-making systems. Finally, we discuss the multiple ways that justice perceptions are related to the use of digital twins of humans.

Keywords Organisational justice · Fairness · Automated decision-making · Human-computer interaction

P. Hellwig (✉) · G. W. Maier
Work and Organizational Psychology, Bielefeld University,
Universitätsstraße 25, 33615 Bielefeld, Germany
e-mail: paul.hellwig@uni-bielefeld.de

G. W. Maier
e-mail: ao-psychologie@uni-bielefeld.de

CoR-Lab, Bielefeld University, Inspiration 1, 33619 Bielefeld, Germany

I. Gräßler et al. (eds.), *The Digital Twin of Humans*,
https://doi.org/10.1007/978-3-031-26104-6_4

1 Introduction

Consider an employee named Mary: She works as a human resource manager at a business that provides supplies to leading automotive companies. She begins her work week by checking this week's projects, which the company's staff scheduling algorithm has assigned to her. One of these algorithm-assigned projects is personnel selection for a vacant clerk position. The personnel selection process is already in an advanced state. An algorithm has ranked all applicants according to their application documents and standardised, asynchronous video interviews. Mary checks the rankings and advises the algorithm to invite the three best-ranked applicants to the second phase of the application process. Then, there is a knock on the door. It is John, Mary's 09:00 AM appointment. John is a factory worker who wants to discuss the results of his quarterly algorithm-based performance appraisal. John is not happy with his result and feels that the result is unfair, given his improvement over the last few months. Does this sound like the distant future or even science fiction? It is not. As we will see in Sect. 2, automated decision-making (ADM) is a hot topic and various applications of ADM have already been investigated. Furthermore, we address the concept of human digital twins (DTs) and how they are related to ADM in Sect. 2. Through the implementation of ADM, work changes and justice and fairness perceptions arise as important issues. Recognising the importance of justice and fairness perceptions in ADM, we introduce the concepts of justice and fairness in Sect. 3. In Sect. 4, we summarise empirical evidence regarding justice and fairness perceptions in ADM. In Sect. 5, we address the limitations of the reviewed literature and future directions for research. In Sect. 6, we present design implications that describe how to design ADM that is perceived as fair by stakeholders. In Sect. 7, we discuss the relationship between digital twins and justice/fairness perceptions.

2 Artificial Intelligence as (an Emerging) Decision-Maker

According to recent reviews regarding automated decision-making (Langer & Landers, 2021; Robert et al., 2020; Starke et al., 2022) the use of artificial intelligence (AI[1]) for decision-making is a hot topic. The aforementioned applications of algorithms in Mary's work environment have been studied in research: There are studies on the allocation of tasks or shifts (Bentler et al., 2022; Mlekus et al., 2022; Ötting & Maier, 2018), the preselection of applicants through video interviews (Langer et al., 2019), and performance evaluation (Lee, 2018). In the field, algorithms are consulted before judicial or business decisions are made (Burridge, 2017; Courtland, 2018), they are used to manage app-based workers (Möhlmann & Zalmanson, 2017), and send mechanics to metro trains that need to be fixed (Hodson, 2014). Since AI, as

---

[1] Like Langer and Landers (2021), we use artificial intelligence as an umbrella term for human-programmed algorithms (e.g., regression models or expert systems) and more recent examples such as machine learning, neural networks, and deep learning.

a new decision-making authority, disrupts the work context, discussions about the advantages and disadvantages of its use are inevitable. Supporting the use of ADM is the possibility of higher decision quality and efficiency (Langer & Landers, 2021). It has also been found that participants perceive ADM as less biased than human decision-making (HDM; Höddinghaus et al., 2021) and that ADM could lead to more objective treatment (Araujo et al., 2018). However, there are also downsides to ADM. It has also been found that ADM can in fact be biased when the decision-making algorithm is trained with biased data (Leicht-Deobald et al., 2019; Robert et al., 2020). Furthermore, some studies also found that people perceived ADM as particularly unfair (Langer et al., 2019) or as less fair than HDM in various situations (Newman et al., 2020; Wang, 2018). This led to important discussions of the justice and fairness perceptions of ADM (Brockner & Wiesenfeld, 2019; Hellwig et al., 2023; Ötting & Maier, 2018; Robert et al., 2020; Töniges et al., 2017). We join this important discussion about justice and fairness perceptions in ADM in this chapter.

2.1   Digital Twins of Humans and ADM

A digital twin (DT) is a digital representation of an object, a machine, or a process (Stark et al., 2020). A DT consists of a digital master, which contains master data (e.g., basic data about the product geometry or material) and a digital shadow, which contains state and process data (Stark et al., 2020). A DT of a human could therefore be a digital representation of the human in a certain context (e.g., automated task allocation). This DT of a human could include various stable pieces of information, such as their occupation, gender, age, education, training, and competencies (e.g., general mental abilities). Furthermore, it could include self-reports (e.g., work motivation, job satisfaction), third-party reports (e.g., task performance rated by the supervisor, the satisfaction of customers), performance indicators, vital signs (e.g., heart rate), and automatically detected states (e.g., fatigue). The DTs of humans are related to ADM, as they could be used as the basis of ADM. For example, the DTs of humans could be especially useful for adaptive assistance systems that decide about suitable assistance (see chapter Adaptive Assistance Systems: Approaches, Benefits, and Risks by Buchholz and Kopp in this book). Another application scenario could be AI systems that schedule shifts or tasks and consider workers' preferences, work motivation, job satisfaction, and perceptions of work design (Bentler et al., 2022).

3   Justice and Fairness Perceptions

After establishing the relationship between ADM and justice and fairness perceptions in previous research, we now present findings from organisational justice studies that investigated justice and fairness perceptions of decision-making and its consequences. We explain justice and fairness here in greater detail for two reasons. On the one hand, a good understanding of these constructs can be beneficial when designing ADM, because justice emphasises *what* aspects of, for example, a decision process, are important for the people affected by the decisions. On the other hand, the definitions of justice and fairness from justice research will allow us to cluster the literature on ADM and justice/fairness according to different substantive foci in the fourth part of this chapter.

3.1 *Justice—A Definition*

Justice in psychological research is a multilayered construct that has the following four distinct dimensions: procedural, distributive, interpersonal, and informational justice (Colquitt, 2001; Colquitt et al., 2013). The four justice dimensions correspond to the following important steps in decision-making: process (procedural justice), outcome (distributive justice), and interaction and communication when enacting decisions (interpersonal justice and informational justice). These dimensions consist of *justice rules* (Colquitt & Zipay, 2015) that were identified in psychological studies and clustered into a four-dimensional framework (Colquitt, 2001). Perceived adherence to these rules results in the perception that a decision process, an outcome, or the enactment of a decision are appropriate (Colquitt & Zipay, 2015). Research furthermore argues that adherence to these justice rules by decision-making authorities fosters the perception that these authorities are overall fair (Colquitt & Zipay, 2015; Rodell et al., 2017). The justice rules therefore provide a good indication of what people care about in decision-making.

Procedural justice is about the appropriateness of decision procedures and consists of seven rules. A decision procedure is appropriate when people perceive that they are able to express their views and feelings during a procedure (voice), have an influence over the outcome, and are able to appeal the outcome. Additionally, a procedure should be perceived as being applied consistently, free of bias, and based on accurate information, and it should be perceived to uphold ethical and moral standards (Colquitt, 2001; Leventhal, 1980; Thibaut & Walker, 1975).

Distributive justice is about the appropriateness of outcomes and consists of four rules. An outcome is perceived as appropriate when it reflects the effort put into one's work, is appropriate for the completed work, reflects what one has contributed to the organisation, and is justified given one's performance (Colquitt, 2001). These justice rules are based on equity theory (Adams, 1965). According to equity theory (Adams, 1965), people compare their outcomes (e.g., salary) with their invested input (e.g., effort). This output/input ratio is compared to a person's previous ratios or the ratios of other people. According to Deutsch (1975), equity means that a person's 'input to the group outcome will determine his [or her] relative share of it' (Deutsch, 1975, p. 144). The equity principle is often chosen to measure distributive justice because it is best suited for economic contexts, where productivity is the main goal. However, there are other principles that could be perceived as fair in different decision contexts (Deutsch, 1975; Leventhal, 1976). Another such principle is equality. According to this principle, outcomes are equally distributed, regardless of individual investments or need. The goal of allocations according to the equality norm is to foster/maintain social relations. Finally, according to the need principle, greater need results in a greater outcome. The goal of this principle is to foster personal development or welfare.

Interpersonal justice is about perceived respectfulness and propriety in the implementation and enactment of decisions. Interpersonal justice requires decision-making authorities to be perceived as treating people politely, with dignity, and with respect, and these authorities should refrain from improper remarks or comments (Bies & Moang, 1986; Colquitt, 2001; Shapiro et al., 1994).

Informational justice is about the perceived truthfulness and adequacy of explanations. Informational justice requires decision-making authorities to be perceived as being candid in communications, explaining procedures thoroughly, communicating details in a timely manner, and tailoring communications to a person's specific needs. Additionally, explanations regarding procedures must be perceived as reasonable (Bies & Moang, 1986; Colquitt, 2001; Shapiro et al., 1994).

3.2 Consequences of Justice Perceptions

Several meta-analyses show that justice perceptions (of all dimensions) are related to attitudes and behaviour (Cohen-Charash & Spector, 2001; Colquitt et al., 2001, 2013; Skitka et al., 2003; Visweswaran & Ones, 2002) and underline the importance of justice perceptions in work contexts. Justice perceptions are related to important perceptions such as outcome satisfaction, job satisfaction, organisational commitment, and trust in supervisors and the organisation. Justice perceptions furthermore are related to affect. People who perceive procedures, outcomes, explanations, or interpersonal treatment as just report more positive and fewer negative affect. Finally, justice perceptions are positively correlated with performance and organisational citizenship behaviour (OCB; extra role behaviour, such as helping colleagues), and they are negatively correlated with counterproductive work behaviour (CWB).

3.3 Fairness—A Definition

The literature differentiates between justice perceptions and overall fairness perceptions (Ambrose et al., 2015; Ambrose & Schminke, 2009). Whereas justice is 'the perceived adherence to rules that reflect appropriateness in decision contexts', fairness is a holistic (overall) view of appropriateness (Colquitt & Zipay, 2015, p. 188). Therefore, the measures for justice and overall fairness are different. In justice measures, participants are asked to rate different features of, for example, a process—so justice measures indirectly measure fairness (Colquitt, 2001, 2021). Measures that examine perceptions of overall fairness explicitly ask how the fairness of an entity (e.g. an organisation) is perceived—fairness is thus measured directly (Ambrose & Schminke, 2009). Justice and overall fairness perceptions are connected as follows: Justice perceptions lead to overall fairness perceptions (Colquitt & Zipay, 2015). The overall impression that one's boss is fair comes from observations of the boss in previous decision-making situations. A boss that was perceived as adhering to justice rules in previous decisions is likely to be perceived as fair. There is also the concept of facet fairness (Colquitt & Zipay, 2015), which asks about the overall fairness perceptions of a certain justice dimension (e.g., 'How fair was the evaluation process?'). Asking these questions is common in interdisciplinary research (see Sect. 4). In our literature review in Sect. 4 of this chapter, we consider the different measurement approaches. Therefore, we cluster the relevant studies according to whether they measured justice/fairness directly or indirectly.

3.4 Subjectivity of Justice and Fairness Perceptions

The evaluation of justice and fairness is subjective (Mikula, 2005). The extent to which a person perceives an adherence to justice rules or fairness depends on the person's perception of a particular situation. Therefore, it is important to gather the perceptions of stakeholders in decision-making. For example, even if explanations for procedures are provided, it is important to determine how stakeholders perceive these explanations.

4 Current Findings on ADM and Justice/Fairness Perceptions

After introducing justice and fairness in decision-making, we present findings on justice and fairness perceptions of ADM, as well as the consequences of justice perceptions of ADM. The purpose of our literature review is to provide vital insights into the design of fair ADM. We drew the studies discussed in this literature review from three review articles on ADM (Langer & Landers, 2021; Robert et al., 2020; Starke et al., 2022) and from additional literature that we knew through our own research on ADM, as well as from our own publications on ADM and justice/fairness perceptions. A publication had to report measures of justice or fairness perceptions of ADM to be included in the review. We identified 29 publications with a total of 43 studies in our review. We report the findings of our review in the following way: In Sect. 4.1, we present the findings of studies in which justice/fairness was the dependent variable. The independent variables were (a) the form of decision-making (HDM vs. ADM), (b) design factors (e.g., ADM is only used in the screening phase of a personnel selection process), or (c) individual characteristics (e.g., gender, computer literacy). Some of these studies also investigated interaction effects.

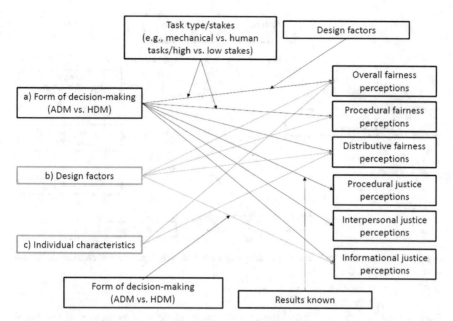

Fig. 1 Overview of investigated relationships in studies in which justice/fairness perceptions were the dependent variable

A typical example of an interaction effect is, for example, whether the positive effect of HDM on the perceived overall fairness is stronger in high-stakes situations (see Fig. 1 for an overview).

In Sect. 4.2, we present studies in which justice perceptions were the independent variable. These studies investigated the effect of justice perceptions (or justice rule adherence) on attitudes, behaviour, and fairness perceptions (see Fig. 2 for an overview). Some moderating variables were also investigated in these studies; for example, one study investigated whether the effect of justice rule adherence has a stronger positive effect on procedural fairness perceptions for women.

4.1   Justice/Fairness as the Dependent Variable

Overall fairness perceptions. Multiple studies investigated the influence of ADM versus HDM on overall fairness perceptions (see Table 1). We included studies that measured fairness as a compound measure of procedural and distributive items or measured fairness in general. The results are mixed. HDM was perceived as fairer in 'human tasks' (tasks that require skills that humans are better at than computers, e.g., subjective judgment; Lee, 2018) such as hiring (Lee, 2018), promotion decisions (Newman et al., 2020), and shift scheduling (Uhde et al., 2020). However, there were also studies in which ADM was perceived to be as fair as HDM. This was

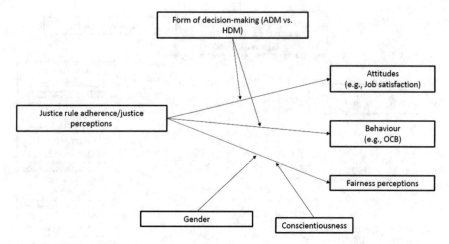

Fig. 2 Overview of investigated relationships in studies in which justice perceptions/justice rule adherence were the independent variable

the case for 'mechanical tasks' (tasks that require skills that computers are better at than humans, e.g., processing quantitative data for objective measures; Lee, 2018) such as work assignment (Lee, 2018) but also in a recruitment context (Suen et al., 2019). In one study, ADM was seen as fairer than HDM in high-impact scenarios involving justice or health decisions (Araujo et al., 2020). Design factors were also related to higher fairness perceptions. Among these design factors were the use of ADM in the screening phase versus in the final decision phase of personnel selection (Hunkenschroer & Lütge, 2021) and the features that were used for automated decisions (Grgić-Hlača et al., 2018). Individual characteristics such as computer literacy (Wang et al., 2020) and political preferences (Grgić-Hlača et al., 2020) also influenced fairness perceptions.

Procedural fairness perceptions. Studies that compared ADM and HDM in terms of procedural fairness also resulted in mixed findings (see Table 2). Some studies found advantages for HDM in terms of procedural fairness perceptions in high-stakes/complex situations (Langer et al., 2019; Nagtegaal, 2021) or in general (Wang, 2018). One study did not find differences between the procedural fairness perceptions of ADM and HDM (Langer et al., 2020). In the study by Nagtegaal (2021), ADM was perceived as more procedurally fair than HDM in low-complexity situations. Design factor transparency (Wang, 2018) influenced perceptions of procedural fairness, but different explanatory styles only if they vary within participants (Binns et al., 2018).

Distributive fairness perceptions. The findings regarding distributive fairness perceptions and ADM were mixed as well (see Table 3). In one study, outcomes that were determined by HDM were perceived as fairer than outcomes that were determined by ADM (Lee & Baykal, 2017). However, the opposite was reported by Howard et al. (2020) and Marcinkowski et al. (2020), who found that ADM was perceived as

Table 1 Overview of studies with overall fairness perceptions as the dependent variable

Study + method + scenario	Independent variables	Results
Lee (2018) 2 vignette studies Various decision-making scenarios	IV1: ADM versus HDM IV2: Task type (human vs. mechanical tasks)	ADM less fair than HDM in human tasks (e.g., hiring), equally fair in mechanical tasks (e.g., work assignment)
Newman et al. (2020) 4 vignette studies, 1 laboratory experiment Various human resource decisions (e.g., promotion)	IV1 (study 1–5): ADM versus HDM IV2 (study 1): decision subject IV2 (study 2): factors considered in decision IV2 (study 5): tansparency (high vs. low)	Study 1–4: ADM less fair than HDM in various situations (main effect) Study 5: ADM less fair than HDM in high transparency condition (interaction effect)
Suen et al. (2019) Online experiments with real applicants Personnel selection	ADM versus HDM	No significant differences in fairness perceptions between ADM and HDM
Araujo et al. (2020) Vignette study Various decision-making scenarios	IV1: ADM versus HDM IV2: Subject of decision: self/other IV3: Impact of decision: high/low IV4: Context: media/health/justice	No significant differences in perceived fairness between ADM and HDM (main effect not significant) When three-way interaction of ADM versus HDM x context x decision impact is taken into account: ADM is perceived as fairer than HDM in the context of justice and health when the decision has a high impact
Hunkenschroer and Lütge (2021) Vignette study Personnel selection	IV1: ADM used in screening phase (vs. final decision phase) of personnel selection IV2: Information: AI reduces human bias IV3: Human oversight versus no human oversight	ADM fairer when used in the screening phase (vs. final decision phase) of personnel selection process and when participants are informed (vs. not informed) that AI reduces human bias

(continued)

Table 1 (continued)

Study + method + scenario	Independent variables	Results
Wang et al. (2020) Online experiment Promotion decision	IV1: Favourable outcome IV2: Biasedness against subgroups IV3: Transparency of ADM IV4: Who built algorithm IV5: Human involvement in decision IV6: Prediction model	Main effects: Algorithm perceived as fair when: Provides favourable outcome, is less biased against subgroups, participant has high (vs. low) computer literacy
	IV7: Education IV8: Computer literacy IV9: Age IV10: Gender IV11: Being from priveleged group	Interactions: A biased algorithm perceived as less fair, when: Built by an outsourced team or transparency is high Stronger negative effect of unfavourable outcome on fairness when: Participant gender is female or education is lower
Grgić-Hlača et al. (2018) Online survey Bail decisions	Feature used for automated bail decisions	Certain features were perceived as fairer than others, e.g., current charges are fairer than criminal history: family and friends Feature is fair when it can be assessed reliably, is relevant for making the decision, causes the outcome, and can be changed by the defendant
Grgić-Hlača et al. (2020) Online survey Bail decisions	IV1: Features used for automated bail decisions IV2: Age IV3: Gender IV4: Political view IV5: Experience with bail decisions	Age and male gender related positively to fairness perceptions Right-leaning participants evaluated all features used for bail decisions as fairer
Uhde et al. (2020) Vignette study Shift scheduling	IV1: ADM versus HDM (system decides vs. 2 workers decide together) varied between participants All varied within participants: IV2: Argument norm (why does a worker want a day off) IV3: Winner (participant vs. co-worker) IV4: Reason (one of three norms as reason for decisions)	Participants perceived system-made decisions as overall less fair than decisions made collaboratively Decisions based on need norm (instead of equality or equity form) were perceived as more fair overall Decisions based on equality norm perceived as fairer than those based on equity norm

Note IV = independent variable

Table 2 Overview of studies with procedural fairness perceptions as the dependent variable

Study + method + scenario	Independent variables	Results
Langer et al. (2020) Online experiment Personnel selection	ADM versus HDM	ADM perceived as equally procedurally fair compared to HDM
Nagtegaal (2021) 2 vignette studies Various decision-making scenarios	IV1: ADM versus HDM IV2: Complexity (low vs. high: e.g., calculating pensions vs. hiring)	ADM perceived as the most procedurally fair in low-complexity decisions HDM perceived as fairer than ADM in high-complexity decisions Combination of ADM and HDM perceived as the most procedurally fair in a high-complexity decision
Langer et al. (2019) Vignette study with videos Personnel selection	IV1: ADM versus HDM IV2: High stakes versus low stakes (personnel selection vs. training)	ADM perceived as less procedurally fair than HDM in high-stakes situations (interaction effect)
Wang (2018) 3 vignette studies Judicial risk assessment	Study 1: IV1: ADM versus HDM IV2: Accuracy of decision-making authority (unknown/high/low) Study 2: IV1: Transparency Study 3: IV1: ADM versus HDM IV2: Disparate impact	Study 1: ADM was perceived as less procedurally fair than HDM Study 2: Transparent algorithms perceived as more procedurally fair than non-transparent algorithms Study 3: HDM perceived as more procedurally fair than ADM
Binns et al. (2018) 2 vignette studies (Study 1: between subjects; Study 2: within subjects) Various decision-making scenarios	IV1: Different explanation styles for automated decisions IV2: Various decision-making scenarios (e.g., applying for a loan, applying for a promotion)	Study 1 (between participants): No explanation style was superior to any other explanation style Study 2 (within participants): Case-based explanation (the decision is explained with a similar case from the database) perceived as less procedurally fair than other explanation styles

Table 3 Overview of studies with distributive fairness perceptions as the dependent variable

Study + method + scenario	Independent variables + scenario	Results
Lee and Baykal (2017)	IV1: ADM versus HDM	Outcome from HDM perceived as fairer than outcomes from ADM
Laboratory experiment	IV2: Interpersonal power (tendency to lead group decisions)	Effect stronger for participants with higher scores of interpersonal power
Allocation of chores	IV3: Computer programming knowledge	More computer programming knowledge led to algorithm-allocated outcomes being perceived as less fair
Howard et al. (2020)	ADM versus HDM	Automatically generated schedule perceived as fairer than manually generated schedule
Field study with two cohorts		
Scheduling of medical trainees		
Marcinkowski et al. (2020)	ADM versus HDM	Fairness ratings of the selection of applicants were higher for ADM
Vignette study (conducted in laboratory)		
Admission to university programs		
Saxena et al. (2019)	Study 1: IV1: Ability of person to pay back a loan	Both studies: Ratio principle was broadly the fairest principle
2 vignette studies	IV2: Three different distribution principles: one person gets everything, ratio, and equal distribution of loan	Both studies: Fairness ratings of the distribution principle depended on the ability of the two people to pay back the loan
Loan distribution between two persons	Study 2: IV1: Ability of person to pay back a loan IV2: Three different distribution principles IV3: Race of person with higher repayment rate	Study 2: Interaction effect of distribution norm and race of person with higher payback rate on perceived fairness

resulting in fairer outcomes than HDM. The study by Saxena et al. (2019) showed the influence of different norms and individual characteristics on fairness perceptions.

Procedural justice perceptions. The studies regarding procedural justice perceptions revealed advantages for ADM and for HDM in certain aspects of procedural justice (see Table 4). ADM was perceived as less biased than HDM (Höddinghaus et al., 2021; Marcinkowski et al., 2020; Schlicker et al., 2021). Furthermore, ADM was seen as more consistent than HDM in some studies (study 2 by Acikgoz et al., 2020; study 1 by Kaibel et al., 2019; Noble et al., 2021); however, there were studies that did not find a difference in consistency between ADM and HDM (study 1 by Acikgoz et al., 2020; study 2 by Kaibel et al., 2019; Langer et al., 2019). HDM had advantages for concepts related to perceptions of voice (Höddinghaus et al., 2021; Schlicker et al., 2021) and control (Langer et al., 2019). Furthermore, HDM was related to perceptions that there was more of an opportunity to perform in some studies (study 1 by Kaibel et al., 2019; Acikgoz et al., 2020; Noble et al., 2021), but this was not found in other studies (study 2 by Kaibel et al., 2019).

Interpersonal justice perceptions. ADM was rated as less interpersonal just in all included studies (see Table 5).

Informational justice perceptions. The findings regarding informational justice are mixed. ADM had advantages in two studies, but it was perceived as being less informational just in another study (see Table 6).

4.2 Justice Rule Adherence/Justice Perceptions as the Independent Variable

Studies in which justice perceptions were the independent variable mostly investigated the effect of the perceived procedural justice on various outcomes (see Table 7). In general, positive procedural justice perceptions in ADM resulted in positive reactions, such as higher job satisfaction, organisational commitment, cooperation, and trust in the decision-making authority, and more OCB (Hellwig & Maier, 2021; Ötting, 2021; Ötting & Maier, 2018). Furthermore, perceptions of voice (Hellwig et al., 2023) resulted in higher overall fairness perceptions. In Dineen et al. (2004), procedural justice rule adherence resulted in higher procedural fairness perceptions. This study furthermore showed that the informational justice rule 'timeliness' is also related to procedural fairness perceptions.

4.3 Conclusion of the Literature Review

In terms of the perceptions of overall fairness and procedural fairness, ADM leads to negative reactions in some scenarios. These scenarios revolved around 'human tasks', 'complex decisions', or 'high stakes'. A common scenario in all of these studies was personnel selection/hiring. Consistent with these results, participants reported higher perceptions of fairness when the influence of ADM on the final decision in personnel selection was restricted. ADM that was only used in the screening phase was perceived as fairer than ADM that was used in the final decision phase. ADM was also found to be less fair than HDM in judicial risk assessment.

Table 4 Overview of studies with procedural justice perceptions as the dependent variable

Study + method + scenario	Independent variables	Results
Höddinghaus et al. (2021)	IV1: ADM versus HDM	ADM perceived as higher in integrity than HDM (unbiasedness, incorruptibility, and honesty)
Vignette study	IV2: Decision subject (disciplinary vs. mentoring)	HDM perceived as higher in benevolence than ADM (how much authority considers interests, needs, and preferences)
Various decision-making scenarios		
Marcinkowski et al. (2020)	ADM versus HDM (within manipulation)	ADM perceived as less biased than HDM
Vignette study (conducted in laboratory)		
Admission to university programs		
Schlicker et al. (2021)	IV1: ADM versus HDM	ADM perceived as less biased than HDM
Vignette study	IV2: Provision of explanation (no explanation vs. equality explanation vs. equity explanation)	HDM perceived as offering participants more voice
Shift scheduling		
Acikgoz et al. (2020)	IV1: ADM versus HDM	Study 1: ADM and HDM perceived as equally consistent and job-related, HDM perceived as offering more of a chance to perform
2 vignette studies	IV2: Outcome favourability (accept/reject)	Study 2: ADM perceived as more consistent than HDM HDM perceived as more job-related and as offering more of a chance to perform
Personnel selection		ADM and HDM equal in perceived reconsideration opportunity
Kaibel et al. (2019)	ADM versus HDM	ADM perceived as more consistent than HDM in vignette study, but not in online experiment
Vignette study Online experiment Personnel selection		HDM perceived as offering more of an opportunity to perform in vignette study, but not in online experiment

(continued)

Table 4 (continued)

Study + method + scenario	Independent variables	Results
Noble et al. (2021)	IV1: ADM versus HDM	HDM perceived as more job-related, as offering more of an opportunity to perform, and as offering more reconsideration opportunities than ADM
Vignette study	IV2: Outcome favourability (acceptance/rejection/unknown)	ADM perceived as more consistent than HDM
Personnel selection		Effect of HDM on perceptions of opportunity to perform and reconsideration opportunity stronger in conditions in which the result is unknown
Langer et al. (2019)	IV1: ADM versus HDM	No differences in perceived consistency between ADM and HDM
Vignette study with videos	IV2: High stakes versus low stakes (personnel selection vs. training)	
Personnel selection		HDM perceived as offering more control over situation in high-stakes contexts
Langer et al. (2020)	ADM versus HDM	HDM perceived as offering more of an opportunity to perform
Online experiment		
Personnel selection		No differences in perceived job-relatedness between ADM and HDM
Wang (2018)	Study 1: IV1: ADM versus HDM IV2: Accuracy of decision-making authority (unknown/high/low)	Study 1: HDM perceived as more accurate than ADM in conditions in which accuracy was unknown
3 vignette studies		
Personnel selection		
Gonzalez et al. (2019)	IV1: ADM versus HDM	No differences in procedural justice perceptions between ADM and HDM
Vignette study	IV2: Outcome favourability (hired / not hired)	
Personnel selection		
Binns et al. (2018)	IV1: Different explanation styles for automated decisions	Study 1 (between subjects): No effect of explanation styles on perceived appropriateness of factors used for decision

(continued)

Table 4 (continued)

Study + method + scenario	Independent variables	Results
2 vignette studies (Study 1: between subjects; Study 2: within subjects) Various decision-making scenarios	IV2: Various decision-making scenarios (e.g., applying for a loan, applying for a promotion)	Study 2 (within subjects): Case-based explanation style (the decision is explained with a similar case from the database) resulted in more negative perceptions of the use of appropriate factors for decision

Table 5 Overview of studies with interpersonal justice perceptions as the dependent variable

Study + method + scenario	Independent variables	Results
Acikgoz et al. (2020) 2 vignette studies Personnel selection	IV1: ADM versus HDM IV2: Outcome favourability (accept/reject)	Study 2: HDM perceived as more polite, respectful, and considerate than ADM
Gonzalez et al. (2019) Vignette study Personnel selection	IV1: ADM versus HDM IV2: Outcome favourability (hired/not hired)	ADM perceived as having less interpersonal justice than HDM
Noble et al. (2021) Vignette study Personnel selection	IV1: ADM versus HDM IV2: Outcome favourability (acceptance/rejection/unknown)	ADM perceived as less polite, respectful, and considerate than HDM, and ADM was perceived as asking less-appropriate questions
Schlicker et al. (2021) Vignette study Shift scheduling	IV1: ADM versus HDM IV2: Provision of explanation (no explanation vs. equality explanation vs. equity explanation)	HDM perceived as having more interpersonal justice than ADM

Table 6 Overview of studies with informational justice perceptions as the dependent variable

Study + method + scenario	Independent variables	Results
Acikgoz et al. (2020) 2 vignette studies Personnel selection	IV1: ADM versus HDM IV2: Outcome favourability (accept/reject)	Study 2: HDM perceived as more open (honest and straightforward) than ADM
Höddinghaus et al. (2021) Vignette study Various decision-making scenarios	IV1: ADM versus HDM IV2: Decision subject (disciplinary vs. mentoring)	ADM perceived as more transparent than HDM
Schlicker et al. (2021) Vignette study Shift scheduling	IV1: ADM versus HDM IV2: Provision of explanation (no explanation vs. equality explanation vs. equity explanation)	ADM perceived as having more informational justice than HDM when no explanation is provided for a decision Provision of explanation had a positive effect on perceived informational justice for HDM but not for ADM

One implication of these results is that companies should proceed with caution when the automation of personnel selection or judicial decisions is planned. A second finding was that ADM and HDM carried unique advantages in terms of justice perceptions: ADM was perceived as less biased than HDM. HDM, on the other hand, was viewed as offering people more voice, control, the opportunity to perform, and more respectful treatment. Furthermore, HDM was associated with higher perceptions of interpersonal justice. These insights can be used to find appropriate application areas for ADM (scenarios where unbiasedness is important, e.g., decisions where people believe that bias has determined their outcomes in the past) and HDM (scenarios where voice, control, the opportunity to perform, and the perceived treatment are important,e.g., personnel selection). The results of studies in which justice was the independent variable showed that perceptions of procedural justice are positively correlated with fairness perceptions and work outcomes. This highlights that it is important to consider justice perceptions in ADM.

Table 7 Overview of studies with justice perceptions as the independent variable

Study + method + scenario	Independent variables	Results
Dineen et al. (2004) Vignette study (within design)	IV1: Features of decision-making process (opportunity to provide additional information, consistent process, opportunity to appeal decision, timely feedback)	All features related to procedural fairness perception HDM perceived as procedurally fairer than ADM Females reacted more positively to HDM, ability to provide additional information, consistency, and ability to appeal decision
Personnel selection	IV2: ADM versus HDM IV3: Gender IV4: Conscientousness Dependent variable (DV): Procedural fairness perceptions	Conscientious participants reacted more positively to HDM, ability to provide additional information, consistency, and ability to appeal decision
Hellwig et al. (2023) Online experiment Task allocation	IV: Voice perceptions (high/low) DV: Overall fairness perceptions	Voice perceptions resulted in higher perceived overall fairness of the system
Hellwig and Maier (2021) Vignette study Task allocation	IV1: Voice perceptions (high/low) IV2: ADM versus HDM DVs: Satisfaction with the decision process, job satisfaction, organisational commitment, OCB, and trust in the decision-making authority	Voice perceptions resulted in more satisfaction with the decision process, job satisfaction, organisational commitment, OCB, and trust in the decision-making authority Positive effect of voice for ADM and HDM
Ötting (2021) 2 vignette studies	IV1: Procedural justice perceptions (high/low) IV2: ADM vs. HDM DVs: Positive affect, negative affect, trust, identification with organisation, job satisfaction, organisational commitment, OCB, cooperation, and CWB	Procedural justice perceptions related positively to positive affect, trust, identification with organisation, job satisfaction, organisational commitment, OCB, and cooperation in both studies Procedural justice perceptions related negatively to negative affect (both studies) and CWB (only study 2)

(continued)

Table 7 (continued)

Study + method + scenario	Independent variables	Results
Ötting and Maier (2018)	IV1: Procedural justice perceptions (high/low) IV2: ADM versus HDM	Procedural justice perceptions resulted in more job satisfaction, organisational commitment, OCB, and cooperation, and less CWB (only study 1)
2 vignette studies		
Task allocation and training allocation	DVs: Job satisfaction, organisational commitment, OCB, cooperation, and CWB	Positive effect of procedural justice for ADM and HDM

5 Limitations and Future Research

5.1 Lack of Studies in Real Interaction Scenarios

Most of the reviewed studies were vignette studies in which participants imagined a certain situation, without any real interaction with ADM. This has some limitations. First, the limitations of a system might be more visible in real interaction scenarios: The participants might question the capabilities of the system in real life, whereas they might not question the capabilities of a system that is described in a vignette. Second, as Langer and Landers (2021) noted in their review on ADM, vignettes investigate systems imagined by researchers, but this might not represent real ADM systems well. Therefore, future research should involve experiments with real interaction scenarios and field studies in which ADM is used.

5.2 Mixed Results Regarding Justice and Fairness Perceptions

Our literature overview indicated mixed results regarding fairness perceptions in ADM. It is not clear how these mixed results should be explained. Langer and Landers (2021) suspected that possible moderators, such as the task in which ADM is used (e.g., personnel selection vs. scheduling), might have caused these inconclusive results. However, there might be an additional explanation for these results. It could be that unbiasedness was very salient in some situations. In addition, it could be that the opportunity to voice views and perceptions of treatment were very salient in other situations. This could have resulted in more positive evaluations of the decision-maker that was more associated with one or the other of these situations. In multiple studies in our literature overview, we found that ADM was perceived as less biased than HDM, but HDM was perceived as offering participants more voice and better treatment than ADM. More research that investigates whether ADM or HDM is perceived as fairer and under what conditions is needed.

5.3 More Research Needed on Consequences and Implementation of Justice Rules

More research is needed to investigate the consequences of perceived justice in ADM. Some initial evidence shows that procedural justice perceptions have positive consequences in ADM. However, some studies discuss a possible interaction effect of decision-making authority and perceived procedural justice (Ötting & Maier, 2018; Hellwig & Maier, 2021). These studies argue that ADM's compliance with procedural justice rules might be less effective because participants might be sceptical that procedural justice has the same instrumental and interpersonal benefits when granted by computers. For example, granting voice might be less effective for ADM than for HDM because participants might doubt that computers are able to use voiced input, which could diminish the instrumental value of voice. Moreover, it is not clear whether voice granted by computers serves an interpersonal purpose by signalling that the decision-making authority is interested in a participant's opinion as an important group member. It could be that voice does not have this function when granted by computers. Even though all included publications do not find that justice has weaker effects when computers are used as decision-making authorities, it might still be important to investigate whether justice has the same effects in ADM as in HDM. Furthermore, more research is needed on how justice rules can be implemented in ADM. While Colquitt (2001) described justice rules that are appropriate for decision-making, further work is needed to examine exactly how these justice rules can be implemented. We attempt to explain this with the following examples.

Procedural justice. The voice rule states that participants should have the perception that they can present their views and feelings regarding the decision to the decision-maker. How can this be achieved in ADM? How should people voice their views and feelings to AI? Should they use free answers, rating scales, or both?

Distributive justice. Distributive justice focuses mainly on the equity norm. What is the right norm (or norm combination) for different ADM decisions? Are different norms appropriate for different decisions? Additionally, do people believe that a computer is capable of understanding what a human needs?

Interpersonal justice. One interpersonal justice rule states that people should feel respected by the decision-making authority. How does an AI need to treat people to make them feel respected? Is it even possible for an AI to be perceived as respectful by a human?

Informational justice. One informational justice rule demands that decisions should be explained in a way that allows people to understand them. How can an AI explain its actions so that people can understand them?

6   Design Implications for ADM

6.1   Consider When (Not) to Use ADM

One important design implication is caution in determining when ADM should be used. The results indicate that caution is advised in particular for human tasks (Lee, 2018), complex decisions (Nagtegaal, 2021), and situations in which participants want control, voice, the opportunity to perform, and respectful treatment. These situations may include selection procedures and various human resource decisions (Newman et al., 2020). We advise against fully automating these decisions. AI was perceived as equally fair in mechanical tasks (Lee, 2018) and in decisions with lower risk, such as situations in which AI gives feedback on video interviews for training purposes (Langer et al., 2019). The results furthermore indicate that AI is believed to be unbiased. AI could be useful for situations or teams in which biased decision-making was an issue previously or is particularly important to the affected stakeholders.

6.2   Design Factors that are Related to Fairness Perceptions

There are two factors that are related to fairness perceptions that should be considered. First, the use of ADM should be limited to the earlier stages of decision processes. When ADM was used in the earlier stages of personnel selection processes, it was perceived as fairer (Hunkenschroer & Lütge, 2021). Therefore, for now the use of ADM could be limited to the earlier phases of a decision process and the final decision could be made by a human. Second, appropriate predictors should be used. The predictors that ADM uses were perceived as fairer when they were directly related to the decision, were under the affected person's control, and when it was possible to measure these predictors reliably (Grgić-Hlača et al., 2018). Therefore, the predictors used by ADM should fulfil these requirements.

6.3   Gather Feedback on How Stakeholders Perceive ADM

As justice and fairness perceptions are subjective, stakeholders should be considered when ADM is implemented. We recommend retrieving feedback regarding justice and fairness perceptions from the stakeholders to ensure that the ADM process that is implemented is perceived as just and fair. The goal of the following list (see Table 8) is to serve as a conversation guide, making it possible to retrieve justice and fairness perceptions from stakeholders and to develop ideas on how to foster ADM that is perceived as just and fair. It is based on the questionnaire by Colquitt (2001) and the idea that justice rules can be used to manage justice and fairness perceptions

Table 8 Conversation guide for investigating perceived justice and fairness in ADM

Procedural justice
On a scale from 1 (not at all) to 5 (very much): How much do you perceive...
(1a) ...that you can voice your views and feelings in the decision processes?
(1b) How could this be improved? / What changes would improve your rating by one point?
(2a) ... that you have an opportunity to influence the final result of the decision?
(2b) How could this be improved?
(3a) ... the process as consistent over people and time?
(3b) How could this be improved?
(4a) ... the system as unbiased?
(4b) How could this be improved?
(5a) ... the system as using only accurate information for the decision?
(5b) How could this be improved?
(6a) ... an opportunity to appeal the decision?
(6b) How could this be improved?
(7a) ... that the system adheres to ethical and moral (and legal) rules?
(7b) How could this be improved?
Procedural fairness
(8a) On a scale from 1 (not at all) to 5 (very much): How fair is the decision-making process overall?
(8b) How could this be improved?
Distributive justice
(9a) According to what principle do you think benefits should be distributed by the system?
(9b) Why?
Distributive justice (when the equity norm is agreed on)
On a scale from 1 (not at all) to 5 (very much): How much do you perceive that ...
(10a) ... the system distributes outcomes that reflect the effort you put into the work?
(10b) How could this be improved?
(11a) ... the system distributes outcomes that are appropriate for the work you completed?
(11b) How could this be improved?
(12a) ... the outcomes reflect what you contributed to the organisation?
(12b) How could this be improved?
(13a) ...the system distributes outcomes that are justified given your performance?
(13b) How could this be improved?
Distributive fairness
(14a) On a scale from 1 (not at all) to 5 (very much): How fair is the outcome you received from the system overall?
(14b) How could this be improved?

(continued)

Table 8 (continued)

Interpersonal justice
On a scale from 1 (not at all) to 5 (very much): How much do you perceive that the system...
(15a) ... treats you politely?
(15b) How could this be improved?
(16a) ... treats you with dignity?
(16b) How could this be improved?
(17a) ...treats you with respect?
(17b) How could this be improved?
(18a) ... does not make inappropriate remarks or comments?
(18b) How could this be improved?
Interpersonal fairness
(19a) On a scale from 1 (not at all) to 5 (very much): How fairly does the system treat you overall?
(19b) How could this be improved?
Informational justice
On a scale from 1 (not at all) to 5 (very much): How much do you perceive that the system...
(20a) ... and its explanations are candid?
(20b) How could this be improved?
(21a) ... explains procedures thoroughly?
(21b) How could this be improved?
(22a) ... uses reasonable explanations for procedures?
(22b) How could this be improved?
(23a) ... explains everything in a timely manner?
(23b) How could this be improved?
(24a) ... provides explanations that are tailored to you (vs. generic)?
(24b) How could this be improved?
Informational fairness
(25a) On a scale from 1 (not at all) to 5 (very much): How fair is the systems' way of providing you with explanations and relevant information?
(25b) How could this be improved?
Overall fairness
(26a) On a scale from 1 (not at all) to 5 (very much): How fair is the system overall?
(26b) How could this be improved?

in organisations (Colquitt & Rodell, 2015). The conversation guide includes rating scales for the perceived justice rule adherence and open follow-up questions that can be used to improve justice perceptions. The guide furthermore includes fairness evaluations. We decided to include these additional questions regarding fairness for the following reason. Although research indicates that justice rule adherence contributes to overall fairness perceptions, there might be additional factors that influence the overall fairness ratings of a decision-making authority (e.g., the charismatic qualities of supervisors; Rodell et al., 2017). Therefore, it seemed reasonable to ask additional questions that capture the relevant fairness aspects of decision-making beyond justice rule adherence.

7 Digital Twins of Humans and Justice/Fairness Perceptions

As the final work task of the day, Mary, the human resource manager from the beginning of this chapter, is asked to prepare a short presentation on the DTs of humans. The board member who gave her this task is especially interested in implementing this technology in the company (which will result in DTs of the employees). In her presentation, Mary highlights three ways in which DTs are linked to perceptions of justice and fairness.

7.1 Importance of Justice Perceptions When Creating and Using DTs

As with ADM in general, justice perceptions are probably relevant when the DTs of humans are implemented and used. The DTs of humans will probably be used in ADM as the basis for decisions. The decision of which data should be included in a DT is one in which procedural justice perceptions are likely to be very important. In particular, the worker being modelled by a DT would probably like to have a say in and a right to appeal the decision concerning what data are included in the DT. In addition, workers could demand thorough and candid explanations of what is modelled in their DT and what the DT is used for (informational justice). Perceptions of appropriate treatment will probably also be important when implementing and using a DT: The concerns of the affected workers should be taken seriously, and the worker should always feel respected and treated with dignity (interpersonal justice). In addition, the consequences of using a DT should be perceived as being based on accepted norms (distributive justice).

7.2 A DT Makes the System Aware of a Stakeholder's Justice and Fairness Perceptions

A DT of a stakeholder's perceptions of justice and fairness could be used in systems that make decisions. This mechanism could serve as a continuous feedback loop for the system's designers or for the system itself. In a very autonomous system, a DT could be used to influence the system to change its decision-making when stakeholders perceive the decision-making as unfair. Therefore, the digital twin of a worker could include different pieces of relatively stable information that are relevant to fairness perceptions. For example, it could include preferences for allocation norms, explanation styles, and procedural features. The digital shadow could include current justice and fairness perceptions, opinions on decisions made by the system, and alternative courses of action preferred by the worker. A DT could raise an ADM system's awareness of its users and would make justice/fairness the central design feature of ADM.

7.3 ADM that Is Perceived as Just/Fair Due to the Use of DTs

The DTs of humans could also provide an opportunity to improve justice/fairness perceptions in ADM. The improvement of justice/fairness perceptions could be achieved by incorporating personal data (e.g., interests, skills, experience, and preferences) from relevant stakeholders into decisions that previously only considered processes and production resources. This inclusion of stakeholders' personal data could improve the perceptions of procedural justice (e.g., control over the outcome, the use of accurate data, and allowing stakeholders to have a voice), especially if the DT includes personal data that stakeholders want to be included in the decision.

8 Conclusion

In this chapter, we discussed ADM as an important topic in research and society. Preliminary research has found that it is beneficial when ADM is perceived as just. Research has also identified the strengths (e.g., ADM is perceived as less biased) and weaknesses (ADM is perceived as offering stakeholders less of a voice, less control, and less of an opportunity to perform, and it is perceived as treating people worse) of ADM. Justice and fairness perceptions could also be important when creating and using DTs. DTs could be used to design human-centred technology that is aware of the justice and fairness perceptions of its users and promotes perceptions of justice and fairness through the consideration of stakeholder states.

Acknowledgements This research was supported by the North-Rhine Westfalian graduate school "Design of Flexible Work Environments—Human-Centric Use of Cyber-Physical Systems in Industry 4.0" (Grant AZ 321-8.03-110- 116443 by the Ministry of Culture and Science of the State of North Rhine-Westphalia).

References

Acikgoz, Y., Davison, K. H., Compagnone, M., & Laske, M. (2020). Justice perceptions of artificial intelligence in selection. *International Journal of Selection and Assessment, 28*(4), 399–416. https://doi.org/10.1111/ijsa.12306

Adams, J. S. (1965). Inequity in social exchange. In L. Berkowitz (Ed.), *Advances in experimental social psychology* (pp. 267–299). Academic Press. https://doi.org/10.1016/s0065-2601(08)60108-2

Ambrose, M. L., & Schminke, M. (2009). The role of overall justice judgments in organizational justice research: A test of mediation. *The Journal of Applied Psychology, 94*(2), 491–500. https://doi.org/10.1037/a0013203

Ambrose, M. L., Wo, D. X. H., & Griffith, M. D. (2015). Overall justice. In R. S. Cropanzano & M. L. Ambrose (Eds.), *The Oxford handbook of justice in the workplace.* Oxford University Press. https://doi.org/10.1093/oxfordhb/9780199981410.013.5

Araujo, T., De Vreese, C., Helberger, N., Kruikemeier, S., van Weert, J., Bol, N., Oberski, D., Pechenizkiy, M., Schaap, G., & Taylor, L., et al. (2018). *Automated decision-making fairness in an AI-driven world: Public perceptions, hopes and concerns.* Digital Communication Methods Lab. http://www.digicomlab.eu/reports/2018_adm_by_ai/

Araujo, T., Helberger, N., Kruikemeier, S., & de Vreese, C. H. (2020). In AI we trust? perceptions about automated decision-making by artificial intelligence. *AI & SOCIETY, 35*(3), 611–623. https://doi.org/10.1007/s00146-019-00931-w

Bentler, D., Gabriel, S., Meyer zu Wendischhoff, D., Bansmann, M., Latos, B., Junker, C., & Maier, G. W. (2022). Gestaltung humanzentrierter Entscheidungen einer künstlichen Intelligenz für Personaleinsatzprozesse produzierender Unternehmen. In Gesellschaft für Arbeitswissenschaft (Ed.), *Technologie und Bildung in hybriden Arbeitswelten.*

Bies, R. J., & Moang, J. F. (1986). Interactional justice: Communication criteria of fairness. In R. J. Lewicki, B. H. Sheppard, & M. H. Bazerman (Eds.), *Research on negotiations in organizations* (pp. 43–55). JAI Press.

Binns, R., van Kleek, M., Veale, M., Lyngs, U., Zhao, J., & Shadbolt, N. (2018). 'It's reducing a human being to a percentage': Perceptions of justice in algorithmic decisions. *Proceedings of the 2018 CHI Conference on Human Factors in Computing Systems,* 1–14. https://doi.org/10.1145/3173574.3173951

Brockner, J., & Wiesenfeld, B. M. (2019). Organizational justice is alive and well and living elsewhere (but not too far away). In E. A. Lind (Ed.), *Social psychology and justice* (pp. 213–242). Routledge. https://doi.org/10.4324/9781003002291-10

Burridge, N. (2017, May 10). Artificial intelligence gets a seat in the boardroom: Hong Kong venture capitalist sees AI running Asian companies within 5 years. *Nikkei Asia.* https://asia.nikkei.com/Business/Artificial-intelligence-gets-a-seat-in-the-boardroom

Cohen-Charash, Y., & Spector, P. E. (2001). The role of justice in organizations: A meta-analysis. *Organizational Behavior and Human Decision Processes, 86*(2), 278–321. https://doi.org/10.1006/obhd.2001.2958

Colquitt, J. A. (2001). On the dimensionality of organizational justice: A construct validation of a measure. *The Journal of Applied Psychology, 86*(3), 386–400. https://doi.org/10.1037/0021-9010.86.3.386

Colquitt, J. A. (2021). My journey with justice: Brainstorming about scholarly influence and longevity. In X.-P. Chen & H. K. Steensma (Eds.), *A journey toward influential scholarship: Insights from leading management scholars* (pp. 124–146). Oxford University Press.

Colquitt, J. A., & Rodell, J. B. (2015). Measuring justice and fairness. In R. S. Cropanzano & M. L. Ambrose (Eds.), *The Oxford handbook of justice in the workplace*. Oxford University Press. https://doi.org/10.1093/oxfordhb/9780199981410.013.8

Colquitt, J. A., & Zipay, K. P. (2015). Justice, fairness, and employee reactions. *Annual Review of Organizational Psychology and Organizational Behavior, 2*, 75–99. https://doi.org/10.1146/annurev-orgpsych-032414-111457

Colquitt, J. A., Conlon, D. E., Wesson, M. J., Porter, C. O., & Ng, K. Y. (2001). Justice at the millennium: A meta-analytic review of 25 years of organizational justice research. *The Journal of Applied Psychology, 86*(3), 425–445. https://doi.org/10.1037/0021-9010.86.3.425

Colquitt, J. A., Scott, B. A., Rodell, J. B., Long, D. M., Zapata, C. P., Conlon, D. E., & Wesson, M. J. (2013). Justice at the millennium, a decade later: A meta-analytic test of social exchange and affect-based perspectives. *Journal of Applied Psychology, 98*(2), 199–236. https://doi.org/10.1037/a0031757

Courtland, R. (2018). Bias detectives: The researchers striving to make algorithms fair. *Nature, 558*(7710), 357–360. https://doi.org/10.1038/d41586-018-05469-3

Deutsch, M. (1975). Equity, equality, and need: What determines which value will be used as the basis of distributive justice? *Journal of Social Issues, 31*(3), 137–149. https://doi.org/10.1111/j.1540-4560.1975.tb01000.x

Dineen, B. R., Noe, R. A., & Wang, C. (2004). Perceived fairness of web-based applicant screening procedures: Weighing the rules of justice and the role of individual differences. *Human Resource Management, 43*(2–3), 127–145. https://doi.org/10.1002/hrm.20011

Gonzalez, M. F., Capman, J. F., Oswald, F. L., Theys, E. R., & Tomczak, D. L. (2019). Where's the I-O? Artificial intelligence and machine learning in talent management systems. *Personnel Assessment and Decisions, 5*(3). https://doi.org/10.25035/pad.2019.03.005

Grgić-Hlača, N., Redmiles, E. M., Gummandi, K. P. & Weller, A. (2018). Human perceptions of fairness in algorithmic decision making: A case study of criminal risk prediction. *Proceedings of the 2018 World Wide Web Conference*, pp. 903–912. https://doi.org/10.1145/3178876.3186138

Grgić-Hlača, N., Weller, A., & Redmiles, E. M. (2020). *Dimensions of diversity in human perceptions of algorithmic fairness*. https://doi.org/10.48550/arXiv.2005.00808

Hellwig, P., & Maier, G. W. (2021, September 22–24). *Deine Meinung zählt: Mitspracherecht bei computerbasierten Entscheidungen im Arbeitsalltag* [Conference presenation] 12. Tagung der Fachgruppen Arbeits-, Organisations- und Wirtschaftspsychologie sowie Ingenieurspsychologie der DGPs, Chemnitz, Germany. https://pub.uni-bielefeld.de/record/2958009

Hellwig, P., Buchholz, V., Maier, G. W., & Kopp, S. (2023). Let the user have a say—voice in automated decision-making. *Computers in Human Behavior, 138*, Article 107446. https://doi.org/10.1016/j.chb.2022.107446

Höddinghaus, M., Sondern, D., & Hertel, G. (2021). The automation of leadership functions: Would people trust decision algorithms? *Computers in Human Behavior, 116*, Article 106635. https://doi.org/10.1016/j.chb.2020.106635

Hodson, H. (2014, July 2). The AI boss that deploys Hong Kong's subway engineers. *New Scientist*. https://www.newscientist.com/article/mg22329764-000-the-ai-boss-that-deploys-hong-kongs-subway-engineers/

Howard, F. M., Gao, C. A., & Sankey, C. (2020). Implementation of an automated scheduling tool improves schedule quality and resident satisfaction. *PLOS ONE, 15*(8), Article e0236952. https://doi.org/10.1371/journal.pone.0236952

Hunkenschroer, A., & Lütge, C. (2021). How to improve fairness perceptions of AI in hiring: The crucial role of positioning and sensitization. *AI Ethics Journal, 2*(2). https://doi.org/10.47289/AIEJ20210716-3

Kaibel, C., Koch-Bayram, I., Biemann, T., & Mühlenbock, M. (2019). Applicant perceptions of hiring algorithms - uniqueness and discrimination experiences as moderators. *Academy of Management Proceedings, 2019*(1). https://doi.org/10.5465/AMBPP.2019.210

Langer, M., & Landers, R. N. (2021). The future of artificial intelligence at work: A review on effects of decision automation and augmentation on workers targeted by algorithms and third-party observers. *Computers in Human Behavior, 123*, Article 106878. https://doi.org/10.1016/j.chb.2021.106878

Langer, M., König, C. J., & Papathanasiou, M. (2019). Highly automated job interviews: Acceptance under the influence of stakes. *International Journal of Selection and Assessment, 27*(3), 217–234. https://doi.org/10.1111/ijsa.12246

Langer, M., König, C. J., & Hemsing, V. (2020). Is anybody listening? the impact of automatically evaluated job interviews on impression management and applicant reactions. *Journal of Managerial Psychology, 35*(4), 271–284. https://doi.org/10.1108/JMP-03-2019-0156, www.emerald.com/insight/content/doi/10.1108/jmp-03-2019-0156/full/pdf

Lee, M. K. (2018). Understanding perception of algorithmic decisions: Fairness, trust, and emotion in response to algorithmic management. *Big Data & Society, 5*(1). https://doi.org/10.1177/2053951718756684

Lee, M. K., & Baykal, S. (2017). Algorithmic mediation in group decisions: Fairness perceptions of algorithmically mediated vs. discussion-based social division. *Proceedings of the 2017 ACM Conference on Computer Supported Cooperative Work and Social Computing*, 1035–1048. https://doi.org/10.1145/2998181.2998230

Leicht-Deobald, U., Busch, T., Schank, C., Weibel, A., Schafheitle, S., Wildhaber, I., & Kasper, G. (2019). The challenges of algorithm-based hr decision-making for personal integrity. *Journal of Business Ethics, 160*(2), 377–392. https://doi.org/10.1007/s10551-019-04204-w

Leventhal, G. S. (1976). The distribution of rewards and resources in groups and organizations. *Advances in Experimental Social Psychology, 9*, 91–131. Academic Press. https://doi.org/10.1016/S0065-2601(08)60059-3

Leventhal, G. S. (1980). What should be done with equity theory? In K. J. Gergen, M. S. Greenberg & R. H. Willis (Eds.), *Social exchange* (pp. 27–55). Springer. https://doi.org/10.1007/978-1-4613-3087-5_2

Marcinkowski, F., Kieslich, K., Starke, C., & Lünich, M. (2020). Implications of AI (un-)fairness in higher education admissions: The effects of perceived AI (un-)fairness on exit, voice and organizational reputation. *Proceedings of the 2020 Conference on Fairness, Accountability, and Transparency*, 122–130. https://doi.org/10.1145/3351095.3372867

Mikula, G. (2005). Some observations and critical thoughts about the present state of justice theory and reserach. In S. W. Gilliland, D. D. Steiner, D. P. Skarlicki & K. van den Bos (Eds.), *What motivates fairness in organizations?* (pp. 197–210). Information Age Publishing.

Mlekus, L., Lehmann, J., & Maier, G. W. (2022). New work situations call for familiar work design methods: Effects and mediating mechanisms of task rotation in a technology-supported workplace. *Frontiers in Psychology, 13*. https://doi.org/10.3389/fpsyg.2022.935952

Möhlmann, M., & Zalmanson, L. (2017). Hands on the wheel: Navigating algorithmic management and uber drivers' autonomy. *Procedings of the International Conference on Information Systems (ICIS 2017)*. https://aisel.aisnet.org/icis2017/DigitalPlatforms/Presentations/3

Nagtegaal, R. (2021). The impact of using algorithms for managerial decisions on public employees' procedural justice. *Government Information Quarterly, 38*(1), Article 101536. https://doi.org/10.1016/j.giq.2020.101536

Newman, D. T., Fast, N. J., & Harmon, D. J. (2020). When eliminating bias isn't fair: Algorithmic reductionism and procedural justice in human resource decisions. *Organizational Behavior and Human Decision Processes, 160*, 149–167. https://doi.org/10.1016/j.obhdp.2020.03.008

Noble, S. M., Foster, L. L., & Craig, S. B. (2021). The procedural and interpersonal justice of automated application and resume screening. *International Journal of Selection and Assessment, 29*(2), 139–153. https://doi.org/10.1111/ijsa.12320

Ötting, S. K. (2021). *Artificial intelligence as colleague and supervisor: Successful and fair interactions between intelligent technologies and employees at work.* [Doctoral Dissertation, Bielefeld University, Germany]. https://doi.org/10.4119/unibi/2953489

Ötting, S. K., & Maier, G. W. (2018). The importance of procedural justice in human-machine interactions: Intelligent systems as new decision agents in organizations. *Computers in Human Behavior, 89*, 27–39. https://doi.org/10.1016/j.chb.2018.07.022

Robert, L. P., Pierce, C., Marquis, L., Kim, S., & Alahmad, R. (2020). Designing fair AI for managing employees in organizations: A review, critique, and design agenda. *Human-Computer Interaction, 35*(5–6), 545–575. https://doi.org/10.1080/07370024.2020.1735391

Rodell, J. B., Colquitt, J. A., & Baer, M. D. (2017). Is adhering to justice rules enough? the role of charismatic qualities in perceptions of supervisors' overall fairness. *Organizational Behavior and Human Decision Processes, 140*, 14–28. https://doi.org/10.1016/j.obhdp.2017.03.001

Saxena, N. A., Huang, K., DeFilippis, E., Radanovic, G., Parkes, D. C., & Liu, Y. (2019). How do fairness definitions fare? examining public attitudes towards algorithmic definitions of fairness. *Proceedings of the 2019 AAAI/ACM Conference on AI, Ethics, and Society, 99*–106. https://doi.org/10.1145/3306618.3314248

Schlicker, N., Langer, M., Ötting, S. K., Baum, K., König, C. J., & Wallach, D. (2021). What to expect from opening up 'black boxes'? comparing perceptions of justice between human and automated agents. *Computers in Human Behavior, 122*, Article 106837. https://doi.org/10.1016/j.chb.2021.106837

Shapiro, D. L., Buttner, E. H., & Barry, B. (1994). Explanations: What factors enhance their perceived adequacy? *Organizational Behavior and Human Decision Processes, 58*(3), 346–368. https://doi.org/10.1006/obhd.1994.1041. https://www.sciencedirect.com/science/article/pii/S0749597884710417

Skitka, L. J., Winquist, J., & Hutchinson, S. (2003). Are outcome fairness and outcome favorability distinguishable psychological constructs? a meta-analytic review. *Social Justice Research, 16*(4), 309–341. https://doi.org/10.1023/A:1026336131206

Stark, R., Anderl, R., Thoben, K. D., & Wartzack, S. (2020). WiGeP-Positionspapier: Digitaler Zwilling. *Zeitschrift für wirtschaftlichen Fabrikbetrieb, 115*, 47–50. https://doi.org/10.3139/104.112311

Starke, C., Baleis, J., Keller, B., & Marcinkowski, F. (2022). Fairness perceptions of algorithmic decision-making: A systematic review of the empirical literature. *Big Data & Society 9*(2). https://doi.org/10.1177/20539517221115189

Suen, H. Y., Chen, M. Y. C., & Lu, S. H. (2019). Does the use of synchrony and artificial intelligence in video interviews affect interview ratings and applicant attitudes? *Computers in Human Behavior, 98*, 93–101. https://doi.org/10.1016/j.chb.2019.04.012. www.sciencedirect.com/science/article/pii/S0747563219301529

Thibaut, J. W., & Walker, L. (1975). *Procedural justice: A psychological analysis.* L. Erlbaum Associates.

Töniges, T., Ötting, S. K., Wrede, B., Maier, G. W., & Sagerer, G. (2017). An emerging decision authority: Adaptive cyber-physical system design for fair human-machine interaction and decision processes. In H. Song, D. B. Rawat, S. Jeschke & C. Brecher (Eds.), *Cyber-physical systems* (pp. 419–430). Academic Press. https://doi.org/10.1016/B978-0-12-803801-7.00026-2

Uhde, A., Schlicker, N., Wallach, D. P., & Hassenzahl, M. (2020). Fairness and decision-making in collaborative shift scheduling systems. *Proceedings of the 2020 CHI Conference on Human Factors in Computing Systems, 1*–13. https://doi.org/10.1145/3313831.3376656

Viswesvaran, C., & Ones, D. S. (2002). Examining the construct of organizational justice: A meta-analytic evaluation of relations with work attitudes and behaviors. *Journal of Business Ethics, 38*(3), 193–203. https://doi.org/10.1023/A:1015820708345

Wang, A. J. (2018). Procedural justice and risk-assessment algorithms. *SSRN Electronic Journal*. https://doi.org/10.2139/ssrn.3170136

Wang, R., Harper, F. M., & Zhu, H. (2020). Factors influencing perceived fairness in algorithmic decision-making. *Proceedings of the 2020 CHI Conference on Human Factors in Computing Systems,* 1–14. https://doi.org/10.1145/3313831.3376813

Graph-Theoretical Models for the Analysis and Design of Socio-Technical Networks

Chiara Cappello and Eckhard Steffen ⓘ

Abstract In the recent past, negative relationships between actors in social networks have increasingly been evaluated. In addition to the consideration of the roles of individual actors, the focus of the investigations is also on the properties of the entire network. For this purpose, networks are modelled as signed graphs; this is a graph-theoretic approach that was introduced in the context of social psychological studies on structural balance in networks. Socio-technical networks in the work environment can be represented by signed graphs. We highlight some recent mathematical approaches that can be used for the in-depth analysis and design of socio-technical networks. Indeed, these tools not only allow us to understand the weak and strong points of a network but also show the value of specific relationships and actors. This analysis can then reveal how to improve the efficiency of the network with the minimum effort and, eventually, how to build teams for specific tasks.

Keywords Graph-theoretical models · Socio-technical networks · Signed graphs

1 Introduction

1.1 Motivation

The ongoing digitalisation of work processes is leading to complex socio-technical networks in which people, machines, and even products are networked with one another. In the model-based description of processes, in addition to digital images of machines and products, digital images of the characteristic properties of a human being, the so-called digital twins of humans, which are required for this process, are

C. Cappello (✉) · E. Steffen
Department of Mathematics, Paderborn University,
Warburger Str. 100, 33098 Paderborn, Germany
e-mail: chiara.cappello@uni-paderborn.de

E. Steffen
e-mail: es@uni-paderborn.de

© The Author(s), under exclusive license to Springer Nature Switzerland AG 2023
I. Gräßler et al. (eds.), *The Digital Twin of Humans*,
https://doi.org/10.1007/978-3-031-26104-6_5

93

also created. A more detailed discussion on this topic can be found in Engels (2020). Of course, human beings cannot be summarised by a list of properties. Furthermore, not all properties of a human being may need to be considered. Hence, one of the main issues is to understand which properties are relevant and how the corresponding data can be acquired. In all scenarios in the work environment, it is not just the machine but also the human being that is relevant. Understanding the interactions between all entities in the work environment and implementing these interactions in a socio-technical system allows time-optimised operations in the environment. Thus, the relationships between a human being and the other entities of the network can be seen as an integral part of the human being's digital twin.

Work is a fundamental part of almost everyone's life. While companies always aim to maximise their efficiency and optimise their outcomes, it is imperative to preserve the well-being of the workers. For this purpose, understanding the structure of the network is crucial. For example, a lack of communication may reduce the efficiency of an organisation, or an isolated worker may feel alone and unhappy with their work. Moreover, knowing the actors that play central roles in a working group may help to bring necessary changes into the network, while identifying subgroups could indicate how to build new teams for specific tasks (Valente, 2012; Zhou et al., 2009; Sauer & Kauffeld, 2016). Adjustments in the task flow or small interventions are sometimes enough to improve the network significantly, so understanding the different kinds of relationships inside the company becomes important (de Jong et al., 2014).

Social network analysis is a well-established scientific discipline (Wasserman et al., 1994). Socio-technical networks, a specific kind of social network, are of increasing interest in the research community. Furthermore, with the transition to socio-technical networks or systems, not only the pure analysis but also the design of such networks has increasingly become the focus of scientific investigations. In this chapter, we show how the information given by the graph analysis yields data that are relevant to the digital twin.

This data acquisition can be done with standard graph computations. Since this information reveals the strong and weak points of the graph, it shows the influence of the network on the actor and vice versa. This implies that the network provides additional information on the actor while also indicating whether the position of the actor in their work environment is improvable and, if not, how it can be changed in order to increase both the efficiency of the network and the satisfaction of the actor.

In order to keep the focus of the chapter on the socio-technical aspect of this problem, we only give a short introduction to the mathematical tools that we developed in our research. For deeper insights, see Cappello and Steffen (2022a, 2022b).

1.2 Graph Representation of Socio-Technical Networks

The clearest representation of a social network is given by graphs. A graph G consists of two sets V and E, where E is a set of 2-element subsets of V. The set V is the set of the vertices of G, and E is the set of the edges of G. Instead of writing $e = \{x, y\}$, we write $e = xy$. We say that e is incident to x (and to y) and that x and y are adjacent. In the context of networks, the vertex set V represents the actors and the edge set E represents relations between actors. More informally, if $e = xy$ is an edge, then we say that x and y are linked or connected (by e) and that x and y are neighbours. The number of edges that are incident to a vertex v is the degree of v, which is denoted by $d_G(v)$.

Given a social network, the resulting graph can provide a reliable picture of the relation/communication flow and system. A crucial part of this is the definition of the precise meaning of an edge connecting two actors in the network. If this definition is clearly described, then the analysis of the corresponding graph reveals important properties of the related network. Some of these properties involve the structural characteristics of the entire network, like the connectivity, the density of the graphs, or whether there exist particular subgraphs. Others reveal individual characteristics, like the centrality of one actor, the subgroup this actor belongs to, whether this actor plays any strategical role in the network, etc. (see, e.g., Borgatti and Everett (2006, 2020)). These properties provide not only an understanding of the network but also an understanding of its strong and weak points, as well as of its key actors. This information can lead to many benefits. In particular, once we understand the properties related to the positive outcomes of the networks, we can intervene in order to improve the outcomes. Of course, they depend on the kind of network and the relations that are considered. Evidently, a network representing collaborations among companies will be different from a network representing friendships in a class or a network showing alliances among countries.

We briefly summarise and then analyse some results provided by studies on social networks in the work environment from a mathematical point of view. These considerations play a crucial role in two ways. First, they allow us to describe some target structures that maximise the efficiency of a network and the values of parameters that are related to workers' satisfaction and good outcomes. This implies that, given a certain network, one can change the values of these parameters in order to improve the network. One main concern is related to the possibility of truly changing the network. Some works focusing on this theme (see Valente (2012), Zhou et al. (2009), Sauer and Kauffeld (2016)) showed that these interventions may often be easier than expected. An easily observable case is a connection. Intuitively, it can be said that a bad connection in the communication flow may cause issues; an employee that has no relationships with the other employees may feel unhappy. It has been observed that, in most cases, communication can be increased significantly just by pushing someone to ask a question or, in the case of online communication, by using bots to encourage people to share their opinions (Mazzoni, 2005; Valente, 2012).

Second, vertices, together with how they are embedded in a network, can be seen as digital twins of the workers. Defining a human digital twin is not as easy as defining the digital twin of a machine since not everything can be properly described; many characteristics change over short time frames, and data cannot actually describe all aspects of a real person. By limiting our focus to a specific environment, more can be accomplished. Indeed, employee satisfaction and network outcomes can be seen as consequences of specific values of the graph parameters. Parameters related to

single vertices (like the centrality, degree, etc.) can be directly registered as data for
the digital twins, while parameters related to the entire network can be related to
single vertices by considering how strongly the removal of the vertex would impact
these values. This approach provides a natural way of collecting data. Furthermore,
if changes inside the networks were made using bots, it would be easier to provide
the appropriate characteristics to the bot (e.g., whom it should talk to, whom it should
encourage to talk, etc.).

We focus on signed graphs, which we introduce in Sect. 2. Signed graphs make
it possible to model two types of edges, positive edges and negative edges. The
importance of negative edges has been taken into account more frequently over the
last few years. Harrigan et al. (2019) found that, in the period from 2016–2019, 40,000
articles per year containing the expression 'social network' were published, while
only 550 articles per year contained the expressions 'negative ties' or 'signed graphs'.
Since the year 2000, the number of articles about negative ties and signed graphs has
increased considerably: Indeed, the number of annual publications has doubled every
five years for the last 20 years. A few surveys on signed networks now exist. Tang
et al. (2016) provided a survey on signed network mining in social media and Zheng
et al. (2015) provided a survey on social balance in signed networks. Balance theory,
which is briefly introduced in Sect. 2, is a major field in the analysis of signed social
networks. In Sect. 3, we introduce some recent mathematical concepts that can be
used for the in-depth analysis and design of small social networks; in particular, these
concepts can be used for the detection of actors relevant to the network outcomes
and to determine how they are embedded and clustered in the network. Section 4
summarises our main points and describes implications for further research.

2 Signed Graphs and Balance Theory

2.1 Signed Graphs and Network Properties

Relationships are complex and it often may not be enough to consider only their
existence. In order to address this issue, edges can be supplied with certain properties.
For example, the intensity of a relation may matter (e.g., trust or friendship), so we
can assign a weight to each edge to describe this intensity. Another common situation
is given by unidirectional relations: An actor A may trust another actor B, but this
does not imply that B trusts A. In this case, directions are given to the edges. Our main
focus is on divalent relations, i.e., relations that can be represented as negative or
positive: friendship/enmity, trust/distrust, etc. In order to represent networks with this
kind of relation, we use signed graphs. A signed graph (G, σ) is a graph $G = (V, E)$
together with a function σ from the edge set E to $\{+, -\}$. An edge e is negative if
$\sigma(e) = -$, and N_σ is the set of negative edges of (G, σ). The set $E \setminus N_\sigma$ is the set of
positive edges. The function σ is also called the signature of (G, σ). Furthermore,
the graph $G = (V, E)$ is called the underlying graph of (G, σ). For each vertex v, the

positive degree of v, $d^+(v)$, is the number of positive edges incident to v. Similarly, the negative degree of v, $d^-(v)$, is the number of negative edges incident to v.

The introduction of signed graphs by Harary (1953) was motivated by the development of a formal basis for the psychological theory of social balance of Cartwright and Harary (1956). Balance theory considers tensions caused by the presence of affective relations. It is based on the idea that actors tend toward a certain kind of stability. As a consequence, unstable networks are more prone to change over time to achieve a more balanced state. In particular, stability (and then balance) is given by graphs where, for each actor A, all the enemies of A's friends are also A's enemies, and all the enemies of A's enemies are A's friends. A standard model is given by a network consisting of three actors A, B, and C; each of them has relationships with both of the other actors. If they are all friends, the network will be balanced, since the actors have no reason to change, unless some external situation occurs. A balanced network is also provided in the case in which exactly one relation is good, say the one between A and B, and all the others are bad. This means that A and B are friends and share a common enemy. In other words, the balanced case can be described with the sentences 'the enemy of my enemy is my friend' and 'the friend of my enemy is my enemy' (see Lee et al. (1994)) (Fig. 1).

On the other side, assume that A and B are enemies but that they are both friends with C. C may feel uncomfortable with this situation since his two friends have a bad relationship. In order to solve this tension, either C has to become the enemy of one of his friends, or A and B have to become friends. Similarly, if A, B, and C are all enemies, it may be good for two of them to establish an alliance against the other, since they share a common enemy, so that their negative relation becomes positive. Both of these structures are unstable, so they are called unbalanced. In this case, either 'the friend of my enemy is my friend' or 'the enemy of my enemy is my enemy'.

Harary (1953) concluded that the key structures that can cause an unbalanced situation are circuits. A circuit is a graph $C = (V, E)$ such that the set of its vertices is $V = \{v_1, \dots, v_n\}$ and the set of its edges is $E = \{v_i v_{i+1} : i = 1, \dots, n-1\} \cup \{v_n v_1\}$. In particular, a circuit has length n if it has n vertices. Given a signed circuit $(\mathcal{C}, \sigma)$, its sign is given by the multiplication of the signs of all its edges. This implies that the circuit is negative when it has an odd number of negative edges, and it is positive when the number of negative edges is even. A circuit is balanced when it is

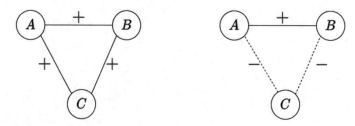

Fig. 1 The two possible balanced triangles

positive; otherwise, it is unbalanced. By observing the previously described cases, we note that A, B, and C form circuits of length 3. In the first two cases, there is an even number of negative edges (0 and 2), and the network is indeed balanced. In the last two cases, the number of negative edges is odd (1 and 3), and therefore the network is unbalanced. This notion can be generalised to all graphs. A graph is balanced if and only if all of its circuits are balanced.

An equivalent definition for balance considers subgroups: A signed graph is balanced if its vertices can be subdivided into two disjoint sets, one of which might be empty, such that all the positive edges connect vertices in the same set, and all the negative edges connect vertices in different sets. Thus, a balanced signed graph (G, σ) has a specific structure with regard to N_σ. In particular, balance is a property of the entire network. By this definition, for each actor A, all the enemies of A's friends are either A's enemies, or they have no relationship with A.

There are $2^{(m-n+1)}$ non-equivalent signatures on a connected graph G with m edges and n vertices. Thus, in general, we cannot expect a large signed network to be balanced, and it is much more common to find a source of tension due to an unbalanced situation. There are simple and fast algorithms for determining whether a signed graph is balanced; see, e.g., Harary and Kabell (1980). Thus, the next natural question is the following: 'How far is a signed graph from being balanced?' To answer this question, observe that, in the example in Fig. 2, the removal of one edge makes the graph balanced. However, there may be more unbalanced circuits in a graph. Sometimes an edge must be removed from each circuit to make the graph balanced, while it may also be that the removal of one particular edge suffices to make the graph balanced, as shown in Fig. 3.

The distance of a signed graph (G, σ) from being balanced is measured by its frustration index (sometimes also called its linear index), that is, the minimum number of edges that have to be removed from G to make the graph balanced. It is denoted by $l(G, \sigma)$ and can be equivalently defined as the minimum number of edges to negate (i.e., to change the sign of) in order to make the graph balanced. The frustration index describes a network property. Determining the frustration index of a signed graph is difficult (see Barahona (1982)).

We underline two points: First, the removal of all negative edges is not always necessary. For example, in the triangle in Fig. 2 with all negative edges, the removal of one edge is enough to reach a balanced state. Second, it may also be necessary to

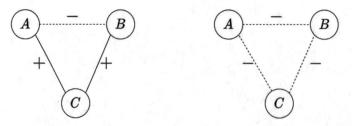

Fig. 2 The two possible unbalanced triangles

Fig. 3 A graph with two
unbalanced circuits and a
frustration index of 1

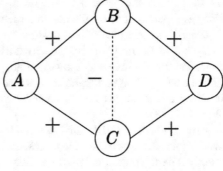

Fig. 4 The removal of the
positive edge *BC* leads to a
balanced graph

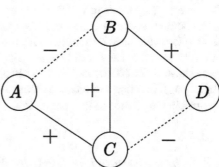

remove positive edges: In Fig. 4, the removal of one negative edge is not enough to obtain balance, but if we delete (or negate) the positive edge *BC*, the graph becomes balanced.

These observations imply that not all relations (bad or good) have the same relevance: Some of them may be harmless and may not create tension, while others may cause instability in many circuits at the same time. Computing the frustration index is not just a way to measure how unbalanced a graph is; it also defines the set of the edges that are the main sources of unbalance.

Sytch and Tatarynowicz (2013) moved the focus of balance theory from the individual's cognitive point of view to a more general approach in order to understand the formation of dyadic and triadic structures in signed graphs. Their work considers networks that describe relations among companies over a longer time frame. They state that these networks do not really tend to balance; rather, they are pulled away from unbalance. The network's unbalance has a strong dependence on the individual's perception: If actors do not perceive stress or tension due to the unbalance, they have no reason to change their network or adapt their behaviour (Hummon & Doreian, 2003). Furthermore, these dynamics require time to change. This implies that in bigger networks, the changes predicted by balance theory seem to be less effective on short timescales (Hummon & Doreian, 2003). At the same time, the behaviour and development of macro-structures strongly depend on the emergence of micro-structures and the relations among them (Stadtfeld et al., 2020). In this way, the strong impact of balance theory on the micro-structures (and on their relations) also has repercussions for the macro-structures.

Clearly, the interpretation of positive and negative edges is context-sensitive and it depends on the properties of the networks under consideration. Therefore, it is not canonical that positive edges always model a good relationship between two actors or that a balanced signed graph always models a network that works well. Examples in which this is not the case are provided by Leskovec et al. (2010) and Aref and Wilson (2018).

Aref and Wilson (2018) analysed the relations among countries during the Cold War. In this case, a positive edge represents an alliance and a negative edge represents enmity. The frustration index of this network is larger than expected. In particular, there were many triads in which a country was allied with two other countries that were enemies. As a consequence, in the case of an actual war, it would not have been clear who would have been allied with whom. This uncertainty may have been one of the reasons that the war did not actually begin.

Leskovec et al. (2010) analysed signed networks in social media. They considered slightly different types of balance theory and evaluated the quality of these balance theories in the context of link prediction.

2.2 The Importance of Negative Edges in Networks

As mentioned in the introduction, the importance of negative edges has been taken into account in network analysis more and more over the last few years. Negative edges allow a more sophisticated analysis of networks since positive and negative edges naturally coincide with relational rating activities in online networks, such as giving likes and dislikes.

A negative edge's value is not only due to the further information they provide; neglecting negative edges results in wrong or inconsistent results in some cases (Stadtfeld et al., 2020). Indeed, on the one hand, having a good relationship with someone may even have the opposite effect of having a bad relationship with them. On the other hand, having a bad relationship is not the same as having no relationship. This implies that, if the effects of relations matter, we cannot represent negative and positive relations with the same kind of edge, nor can we ignore bad relations (by not drawing them at all). A significant example is provided by studies analysing the relationship between employees' satisfaction and centrality (Brass, 1981; Mossholder et al., 2005), that is, how well-/strongly connected employees are. There are many approaches to computing centrality (Estrada & Rodríguez-Velázquez, 2005; Freeman, 1978). The main idea is that a central actor is well connected to the rest of the network because they have many neighbours (i.e., this actor is directly connected to many other actors, which is called degree centrality) or because they are connected to all other actors by short paths, i.e., they are close to all other actors in the network. While it was expected that a central actor would be more satisfied with his job than less-connected actors, it was shown that there was no relation between centrality and satisfaction. Venkataramani et al. (2013) suggested that this result is influenced by the neglect of negative edges. Indeed, an employee may be connected to many other colleagues through negative relations. This situation could easily be a source of stress for the actor, so his satisfaction is not expected to be high. On the other side, many good relations may have a positive impact on the employee. To confirm their hypothesis, Venkataramani et al. (2013) considered the positive and negative networks separately and analysed the relations among centrality, satisfaction, and attachment to the company. They found that employees' centrality has beneficial effects in the positive network and deleterious effects in the negative network. Furthermore, these effects also have direct consequences for employees' attachment to the company.

The case we just described is not an exception: Negative relations are everywhere. Inside families, schools, or workplaces, negative edges may represent conflict, competitions, arguments, bullying, dislike, etc. Similarly, we also find a huge amount of them in social media (negative comments, dislikes, etc.), as well as in economic and political relations. Another interesting aspect is the effect of these negative edges. A negative edge cannot always be considered simply the opposite of a positive edge, since they have different properties. For example, assume that positive (negative) edges represent trust (distrust). If there is trust between A and B and between B and C, then information may flow from A to C. However, if there is distrust between the actors A and B and between B and C, information will probably not flow from A to C, and vice versa, since B has no reason to tell somebody they do not trust something they do not believe. Similarly, receiving a like from a popular person is not the same as receiving a dislike from a hated person (Halgin et al., 2020). It follows that methods used for the analysis of unsigned networks cannot always be directly extended (Everett & Borgatti, 2014). Some require simple adaptations, and others either cannot be used or have to be redefined. For the first case, we can think of the meaning of the degree: Positive and negative degrees have to be distinguished and they must have different meanings. For example, if edges represent likes and dislikes, a high positive degree implies that the actor has a good reputation, while a high negative degree implies a bad reputation. The second case involves more complex parameters, like some extended approaches for centrality (Everett & Borgatti, 2014).

2.3   The Positive Side of Negative Edges

In general, we have always assumed that negative edges represent negative relations. However, as shown by the example of the Cold War, their existence (and frustration) may in fact have positive effects. In order to properly analyse a network, 'positive' and 'negative' relationships must be clearly defined. Indeed, enmity, disagreement, and distrust are all different kinds of negative relations, and all of them have different consequences. The same holds true for positive relations: Friendship, agreement, and trust may have different effects on an actor's behaviour. For example, we may be good friends with a colleague but, at the same time, we may consider him a bad worker and never ask him for advice at work. In particular, there is a key difference between relations with and without emotions: While good (respectively bad) relations that include good (bad) emotions almost always positively (negatively) affect the network, good (bad) relations that do not include emotions may have negative (positive) consequences.

The first approaches took into account positive and negative relations of the same kind, such as friend/enemy, like/dislike, etc. The main idea was that positive and negative edges are mutually exclusive, so we cannot find both between two actors (Chiaburu & Harrison, 2008; Sherf & Venkataramani, 2015). Over the last few years, the idea of considering two different kinds of edges, like enmity/agreement or friendship/competition, has been emerging (Ingram & Roberts, 2000; Leskovec et al., 2010; Marineau et al., 2018; Methot et al., 2017). In this case, a positive and a negative edge may coexist among two actors.

This perspective is particularly interesting in the work environment. In this case, a key aspect is the trust (distrust) in a coworker's ability, i.e., people the actor would (not) ask for advice if they needed it. Other important relations are friendship/enmity and agreement/disagreement, in particular when some jobs or decisions have to be performed together. Combining these different aspects of relations leads to a different perspective, since the overlapping of a positive and a negative edge implies dissonant connections. Such edges are called parallel, and the set of parallel edges between two vertices is called a multi-edge.

Brennecke (2020) conducted a study on dissonant ties in a knowledge-intensive work environment. In that study, an actor is connected with a positive edge to colleagues whom the actor would ask for advice and with a negative edge to colleagues the actor finds it difficult to work with. Interestingly, it turns out that this kind of multi-edge can benefit the actor and the entire network. First, it is known that actors usually seek advice from people who are considered accessible and easy to interact with (Hofmann et al., 2009). This implies that employees who choose colleagues who are difficult to interact with have access to unique resources, since these resources are not easily shared. Second, asking someone for advice implies being willing to deal with disagreement and different points of view and, at the same time, develops this ability. This characteristic plays a particularly important role in knowledge-intensive work environments (Alvesson, 2004). Lastly, Brennecke (2020) indicated that the tensions caused by negative edges may work as a cognitive catalyst in an employee's search for advice. Brennecke's work shows the existence of a positive relationship between actors' performances and the number of dissonant ties they have. Dissonant ties between two actors produce a particular frustrated graph. Indeed, since we do not consider loops, the 2-multi-edge with one positive and one negative edge is the smallest possible unbalanced graph. Its frustration index is 1, i.e., there is exactly one negative edge, and the removal of any single edge makes the graph balanced. Furthermore, given a network, each dissonant tie increases the frustration index by exactly one. A network that contains dissonant ties cannot be balanced. Recall that a network is balanced if it can be subdivided into two groups (one might be empty) such that negative edges connect only actors belonging to different groups and positive edges connect actors in the same group. The case in which one of these groups is empty is quite a rare situation. In this case, there are no bad relations, so communication is smooth. However, the absence of negative edges prevents the positive outcomes caused by dissonant ties. In the case in which the network is clearly divided into two non-empty parts that do not communicate with each other, actors may have no reason to change the network and to interact with the other group. So, what are the possible scenarios for the unbalanced case when we exclude dissonant ties?

In order to further study our hypothesis, we apply Brennecke's considerations to the unbalanced triangles. Again, there are the two cases described previously. If all the edges are negative, we expect that the network will not work properly, and it may also be a source of dissatisfaction for the employees. Consider now the case described in the first graph of Fig. 2, where A and B have a bad relationship, but C shares advice and information with both of them. In this case, A and B may still exchange their information through C, although in an indirect and less efficient way. If we repeat the same argument for a circuit with four actors such that only two of them share a negative edge, we may expect that information between them will flow even more slowly. This approach suggests that the best network may be given by the graph with the minimum number of negative edges with respect to the frustration index such that the length of the cycles containing the negative edges is minimised.

One more confirmation of the possible positive effects of negative edges comes from Park et al. (2020) and Humphrey et al. (2017). They stated that negative edges may have positive outcomes when they are not related to emotions. In their studies, task conflicts and relationship conflicts are considered. On the one side, relationship conflicts can be deleterious for teamwork, since they hinder communication and the exchange of views (Amason, 1996; De Dreu & Weingart, 2003). Conversely, task conflict is caused by different approaches to a problem's solution or a task's performance (Jehn, 1995). This may benefit the network by avoiding premature consensus and by pushing the workers to see the task from different points of view (Janis & Mann, 1977). At the same time, it may also cause dissent and bad relations (De Dreu, 2008). This network is also analysed in relation to the task flow. On the one hand, a bad relation or a lack of communication between two actors who are not connected in the task flow does not influence the outcome in the same way that it does when they share a task. On the other hand, task flows can be more easily adjusted according to the relations among actors, so a company may decide to adapt their task flows in order to optimise the outcomes. Next, dyadic structures with disagreement, i.e., where a negative edge is present, are considered. Since the two workers are connected in the task flow and they disagree on something, we can assume the presence of a relationship between them. This implies that the two actors are connected either by two negative parallel edges or by one positive and one negative edge. Park et al. (2020) showed that the first case is the most deleterious setting, while the other is the most efficient. Indeed, in the first case, a constructive discussion can be held and a better solution can be found by considering the two different approaches. In the second case, disagreement can only further harm the relations since bad emotions obstruct communication. Observe that the first case provides the unbalanced structure that is described by Brennecke (2020). The second case provides a balanced structure in which the number of negative edges is maximised. As before, it seems to describe the worst-case scenario. Indeed, a further balanced case among two actors is given by the presence of two positive edges, that is, agreement and friendship. The risk of this structure is a lack of innovation or deadlocks, but it is still better than the all-negative dyadic structure, since communication is present. These studies assume a particular importance for companies in which the task flow can be modified and adjusted depending on the needs of the company, but it also underlines the value of good relations inside an organisation. Furthermore, it was also proved that negative relationships are less common in a network with high task interdependence (Labianca et al., 1998).

Rothman et al. (2017) investigated negative edges in order to understand when and why they cause positive outcomes in organisations. They asserted that the key factors are the engagement of the employees and their flexibility on different levels (cognitive, emotional, etc.). Indeed, dealing with a negative edge always implies a sort of stress, so an actor needs motivation and the capacity to react in the right way. Labianca et al. (1998) argued that this capacity also depends on four characteristics of the negative edge, that is, strength, reciprocity, cognition (i.e., actors may be not aware of the negative edge), and social distance among actors.

3   Concepts for an In-Depth Analysis of Signed Networks

In this section, we briefly introduce two concepts for analysing signed networks that may gain importance in the future. Further investigations and especially the development of algorithms are necessary. Frustration-critical signed graphs were studied by Cappello and Steffen (2022a), with a focus on the mathematical properties of these graphs. Colouring signed graphs is a classical topic in graph theory. A survey on recent developments in this field of research was given by Steffen and Vogel (2021). Cappello and Steffen (2022b) provided a general framework for the majority of the concepts in signed graph colouring.

Critical subnetworks

An unbalanced signed graph (G, σ) is critical if the frustration index of every proper subgraph of (G, σ) is smaller than the frustration index of (G, σ). That is, every proper subnetwork is closer to being balanced than (G, σ) itself. Thus, the frustration index can be considered a measure of how unbalanced the network is. Frustration-critical signed graphs were investigated by Cappello and Steffen (2022a). Clearly, every unbalanced signed network contains a frustration-critical subnetwork. The removal of any edge in a critical subnetwork changes the frustration index of the subnetwork. This implies that for a given socio-technical system, the removal of any edge that is not contained in any critical substructure does not change the frustration of the graph. This is highly relevant if the (un)balance of a network is related to its outcomes.

Fig. 5 A colouring of a signed graph with the set $C = \{\pm 1, \pm 2\}$

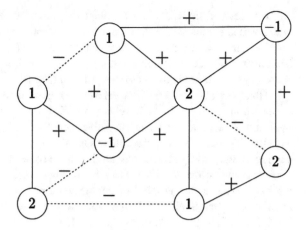

Fig. 6 A colouring of a graph with the set $C = \{1, 2, 3\}$

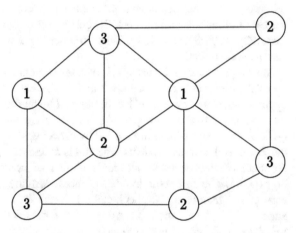

Thus, these actors and their connections have a formative influence on the entire network. For example, the critical subgraphs in the graph shown in Fig. 3 are the circuits A, B, C and B, C, D, and we leave it to the reader to figure out which two edges have to be removed from the graph shown in Fig. 5 to obtain a balanced graph.

Colouring signed graphs

An area of particular interest in graph theory involves colouring. Indeed it is, on the one hand, a hard problem (Garey & Johnson, 1990), but on the other hand, it offers many applications and open questions (Formanowicz & Tanas, 2012).

A colouring of a graph $G = (V, E)$ is a mapping c from the vertex set V into the set of integers $\mathbb{Z}$ such that for each edge $e = vw$, $c(v) \neq c(w)$. For an example, see Fig. 6.

This method is often used for clustering: If a graph can be coloured with n colours, then its vertices can be subdivided into n groups such that there are no edges between vertices of the same group (Liao et al., 2019). For the signed graphs, this approach has to be revised since, also from a clustering point of view, sharing a positive edge with a vertex cannot be the same as sharing a negative edge. A colouring of a signed graph (G, σ) is a map $c : V \to \mathbb{Z}$, such that $c(v) - \sigma(e)c(w) \neq 0$. In particular, if e is a positive edge, then it holds that $c(v) - c(w) \neq 0$, as in the positive case, while if e is negative, it holds that $c(v) + c(w) \neq 0$ (Fig. 5).

At this point, two remarks have to be made regarding the set of colours to use. First, for each colour $c(v) \in \mathbb{Z}$, we have $-c(v) \in \mathbb{Z}$. Thus, we consider symmetric subsets of $\mathbb{Z}$ only. Second, if $e = vw$ is a negative edge, then the constraint is $c(v) \neq -c(w)$. This implies that, if we use a colour x such that $x \neq -x$, then two adjacent vertices sharing a negative edge can be coloured with the same element. This does not happen if we use the colour 0, since $0 = -0$. In other words, negative edges may have a great impact on the number of colours required, but there are colours that can annul this impact. Such colours are called self-inverse elements. Symmetric subsets ($\{-k, \ldots, -1, 0, 1, \ldots, k\}$ or $\{-k, \ldots, -1, 1, \ldots, k\}$) of $\mathbb{Z}$ can contain at most one self-inverse element depending on whether they contain 0. In Fig. 5, for example, the set of colours used is $\{1, -1, 2\}$, but the set of colours that has to be taken is $\{\pm1, \pm2\}$, so $\chi(G, \sigma) = 4$. Similarly, in order to colour (G, σ), with $N_\sigma = E$, two colours are required, e.g., ±1.

We may also use cyclic groups $\mathbb{Z}_k$ as sets of colours. If k is even, then they contain two self-inverse elements, namely 0 and $\frac{k}{2}$. To deal with this issue, different kinds of colouring were developed (see Steffen and Vogel (2021)), and each of them takes into account different numbers of self-inverse elements. In Cappello and Steffen (2022b), these approaches are unified by choosing sets of colours with a predefined number of self-inverse elements. The aim is to find the minimum number of colours required, which is called the chromatic number of the graph and is denoted as $\chi(G)$ or $\chi(G, \sigma)$. The chromatic number of the underlying graph G is just one specific case of a signature of G; specifically, it is the case when there are no negative edges, i.e., $N_\sigma = \emptyset$. The other extremal case occurs when all edges are negative, i.e., $N_\sigma = E$. Then, all vertices can be coloured with the same element, as long as it is not self-inverse, such as, e.g., 1.

The need to use colour pairs may significantly increase the chromatic number of the signed graph with respect to the chromatic number of the underlying graph. In particular, Cappello and Steffen (2022b) found that, if t self-inverse colours are allowed, then $\chi(G, \sigma) \leq 2\chi(G) - t$.

Studies on the clustering of signed graphs already exist. The most common usage consists of the analysis of coalitions (particularly in politics (Aref & Neal, 2020) or in community detection (Zahedinejad et al., 2019)).

The main approach consists of the partition of the graph into groups of vertices such that vertices belonging to the same group do not share negative edges. That is, there are no enemies/bad relations inside a group. These clusterings can be seen as an application of our colouring process. For better usage in this case, we reverse the meanings of the edges by assuming that negative edges represent friendship/similarity, while positive edges represent enmity/dissimilarity. In this way, positive edges are not accepted inside a group. First of all, we begin by assuming that we do not use self-inverse colours, so we have $\pm1, \ldots, \pm n$ as colours. All the sets of vertices coloured with the same element x form a group containing no negative relations/dissimilarity. We call this group the colour class of x. Furthermore, for each colour $x \in C$, the colour class of x can only share positive edges with the colour class of $-x$. In particular, since we are considering negative edges representing positive relations, the graph provided by the colour classes of x and $-x$ is balanced (from a psychological point of view). Observe that if we only aim to minimise the colour classes, different colour classes may also be connected by some negative edges. In order to obtain classes only connected by positive edges, some constraints must be added. Labianca et al. (1998) asserted that friendship across groups is not significantly related to the perception of inter-group conflict, but this is not the case for enmity. This implies that, if positive edges are the majority, negative edges (i.e., good relations) lose some value.

Finally, observe that allowing the usage of self-inverse elements in the colouring means considering the presence of different kinds of actors, that is, actors who only belong to groups in which there are no relations among the vertices.

4 Conclusion

Considering a network in a work environment provides data in two ways: On the one hand, it describes the surroundings and the context in which the actor is working, which of course have a strong impact on their life and satisfaction. On the other hand, it provides information on the role that the actor is playing and on their value inside the network. This kind of data has two main advantages: First, the methods of digital analysis allow us to collect data in a reasonable amount of time. Second, the data reveal information that can be obtained only through in-depth analysis and information that the actor may be unaware of.

These advantages make these data a valid candidate for use in a digital twin. In particular, these data not only make it possible to understand whether the well-being of the actor can be improved by changing their network or by supporting them in some part of their job, but also allow the company to create the best task-oriented environment.

Some tools for this kind of analysis are already used for social research. In this chapter, we suggest two new approaches. One approach uses colouring, which concerns network clustering and is an extension of standard clustering. The other uses the frustration index and the critical structures causing the frustration. The latter approach requires more sophisticated methods, but it also provides a deeper significance for each relation and actor. Since these methods are more technical and may divert the reader from the main focus of the chapter, we only describe a general approach and refer the reader to Cappello and Steffen (2022a) for more specific explanations.

Acknowledgements Chiara Cappello and Eckhard Steffen are members of the research programme 'Design of Flexible Work Environments—Human-Centric Use of Cyber-Physical Systems in Industry 4.0', which is supported by the North Rhine-Westphalian funding scheme 'Forschungskolleg'.

References

Alvesson, M. (2004). *Knowledge work and knowledge-intensive firms*. Oxford: OUP.

Amason, A. C. (1996). Distinguishing the effects of functional and dysfunctional conflict on strategic decision making: Resolving a paradox for top management teams. *Academy of Management Journal, 39*(1), 123–148.

Aref, S., & Neal, Z. (2020). Detecting coalitions by optimally partitioning signed networks of political collaboration. *Scientific Reports, 10*(1), 1–10.

Aref, S., & Wilson, M. C. (2018). Balance and frustration in signed networks. *Journal of Complex Networks, 7*(2), 163–189. https://doi.org/10.1093/comnet/cny015, https://academic.oup.com/comnet/article-pdf/7/2/163/28490360/cny015.pdf

Barahona, F. (1982). On the computational complexity of Ising spin glass models. *Journal of Physics A: Mathematical and General, 15*(10), 3241.

Borgatti, S. P., & Everett, M. G. (2006). A graph-theoretic perspective on centrality. *Social Networks, 28*(4), 466–484. https://doi.org/10.1016/j.socnet.2005.11.005

Borgatti, S. P., & Everett, M. G. (2020). Three perspectives on centrality. *The Oxford handbook of social networks* (p. 334).

Brass, D. J. (1981). Structural relationships, job characteristics, and worker satisfaction and performance. *Administrative Science Quarterly, 26*(3), 331–348.

Brennecke, J. (2020). Dissonant ties in intraorganizational networks: Why individuals seek problem-solving assistance from difficult colleagues. *Academy of Management Journal, 63*(3), 743–778.

Cappello, C., & Steffen, E. (2022a). Frustration-critical signed graphs. *Discrete Applied Mathematics, 322*, 183–193. https://doi.org/10.1016/j.dam.2022.08.010, https://www.sciencedirect.com/science/article/pii/S0166218X22003225

Cappello, C., & Steffen, E. (2022b). Symmetric set coloring of signed graphs. *Annals of Combinatorics, 1–17.* https://doi.org/10.1007/s00026-022-00593-4

Cartwright, D., & Harary, F. (1956). Structural balance: A generalization of Heider's theory. *Psychological Review, 63*(5), 277–93.

Chiaburu, D., & Harrison, D. (2008). Do peers make the place? Conceptual synthesis and meta-analysis of co-workers effects on perceptions, attitudes, OCBs, and performance. *The Journal of Applied Psychology, 93*, 1082–103. https://doi.org/10.1037/0021-9010.93.5.1082

De Dreu, C. K. (2008). The virtue and vice of workplace conflict: Food for (pessimistic) thought. *Journal of Organizational Behavior: The International Journal of Industrial, Occupational and Organizational Psychology and Behavior, 29*(1), 5–18.

De Dreu, C. K., & Weingart, L. R. (2003). Task versus relationship conflict, team performance, and team member satisfaction: A meta-analysis. *Journal of Applied Psychology, 88*(4), 741.

Engels, G. (2020). Der digitale Fußabdruck, Schatten oder Zwilling von Maschinen und Menschen. *Gruppe Interaktion Organisation Zeitschrift für Angewandte Organisationspsychologie (GIO), 51*(3), 363–370.

Estrada, E., & Rodríguez-Velázquez, J. A. (2005). Subgraph centrality in complex networks. *Physical Review E, 71*(5). https://doi.org/10.1103/physreve.71.056103

Everett, M., & Borgatti, S. (2014). Networks containing negative ties. *Social Networks, 38*, 111–120. https://doi.org/10.1016/j.socnet.2014.03.005

Formanowicz, P., & Tanaś, K. (2012). A survey of graph coloring - its types, methods and applications. *Foundations of Computing and Decision Sciences, 37*(3), 223–238. https://doi.org/10.2478/v10209-011-0012-y

Freeman, L. C. (1978). Centrality in social networks conceptual clarification. *Social Networks, 1*(3), 215–239. https://doi.org/10.1016/0378-8733(78)90021-7

Garey, M. R., & Johnson, D. S. (1990). *Computers and intractability: A guide to the theory of NP-completeness*. USA: W. H. Freeman & Co.

Halgin, D. S., Borgatti, S. P., & Huang, Z. (2020). Prismatic effects of negative ties. *Social Networks, 60*, 26–33.

Harary, F. (1953). On the notion of balance of a signed graph. *Michigan Mathematical Journal, 2*(2), 143–146. https://doi.org/10.1307/mmj/1028989917

Harary, F., & Kabell, J. A. (1980). A simple algorithm to detect balance in signed graphs. *Mathematical Social Sciences, 1*(1), 131–136.

Harrigan, N., Labianca, G., & Agneessens, F. (2019). Negative ties and signed graphs research: Stimulating research on dissociative forces in social networks. *Social Networks, 60*. https://doi.org/10.1016/j.socnet.2019.09.004

Hofmann, D., Lei, Z., & Grant, A. (2009). Seeking help in the shadow of doubt: The sensemaking processes underlying how nurses decide whom to ask for advice. *The Journal of Applied Psychology, 94*, 1261–74. https://doi.org/10.1037/a0016557

Hummon, N. P., & Doreian, P. (2003). Some dynamics of social balance processes: Bringing Heider back into balance theory. *Social Networks, 25*(1), 17–49. https://doi.org/10.1016/S0378-8733(02)00019-9

Humphrey, S. E., Aime, F., Cushenbery, L., Hill, A. D., & Fairchild, J. (2017). Team conflict dynamics: Implications of a dyadic view of conflict for team performance. *Organizational Behavior and Human Decision Processes, 142*, 58–70.

Ingram, P., & Roberts, P. W. (2000). Friendships among competitors in the Sydney hotel industry. *American Journal of Sociology, 106*(2), 387–423. https://doi.org/10.1086/316965

Janis, I. L., & Mann, L. (1977). *Decision making: A psychological analysis of conflict, choice, and commitment*. Free Press.

Jehn, K. A. (1995). A multimethod examination of the benefits and detriments of intragroup conflict. *Administrative Science Quarterly*, 256–282.

de Jong, J., Curşeu, P., Leenders, R. (2014). When do bad apples not spoil the barrel? Negative relationships in teams, team performance, and buffering mechanisms. *Journal of Applied Psychology*. https://doi.org/10.1037/a0036284

Labianca, G., Brass, D. J., & Gray, B. (1998). Social networks and perceptions of intergroup conflict: The role of negative relationships and third parties. *Academy of Management Journal, 41*(1), 55–67.

Lee, S. C., Muncaster, R. G., & Zinnes, D. A. (1994). 'The friend of my enemy is my enemy': Modeling triadic international relationships. *Synthese, 100*(3), 333–358.

Leskovec, J., Huttenlocher, D., & Kleinberg, J. (2010). Signed networks in social media. In *Proceedings of the SIGCHI Conference on Human Factors in Computing Systems, Association for Computing Machinery* (pp. 1361–1370). New York, NY, USA, CHI '10. https://doi.org/10.1145/1753326.1753532

Liao, W., Deng, K., & Wang, S. (2019). Community detection based on graph coloring. In *2019 IEEE Symposium Series on Computational Intelligence (SSCI)* (pp. 2114–2118). https://doi.org/10.1109/SSCI44817.2019.9002759

Marineau, J. E., Hood, A. C., et al. (2018). Multiplex conflict: Examining the effects of overlapping task and relationship conflict on advice seeking in organizations. *Journal of Business and Psychology, 33*(5), 595–610.

Mazzoni, E. (2005). La social network analysis a supporto delle interazioni nelle comunit virtuali per la costruzione di conoscenza. *TD, Tecnologie Didattiche, 35*, 54–63.

Methot, J. R., Melwani, S., & Rothman, N. B. (2017). The space between us: A social-functional emotions view of ambivalent and indifferent workplace relationships. *Journal of Management, 43*(6), 1789–1819. https://doi.org/10.1177/0149206316685853

Mossholder, K. W., Settoon, R. P., & Henagan, S. C. (2005). A relational perspective on turnover: Examining structural, attitudinal, and behavioral predictors. *The Academy of Management Journal, 48*(4), 607–618.

Park, S., Mathieu, J. E., & Grosser, T. J. (2020). A network conceptualization of team conflict. *Academy of Management Review, 45*(2), 352–375.

Rothman, N., Pratt, M., Rees, L., & Vogus, T. J. (2017). Understanding the dual nature of ambivalence: Why and when ambivalence leads to good and bad outcomes. *The Academy of Management Annals, 11*, 33–72.

Sauer, N., & Kauffeld, S. (2016). The structure of interaction at meetings: A social network analysis. *Zeitschrift für Arbeits- und Organisationspsychologie, 60*, 33–49. https://doi.org/10.1026/0932-4089/a000201

Sherf, E., & Venkataramani, V. (2015). Friend or foe? The impact of relational ties with comparison others on outcome fairness and satisfaction judgments. *Organizational Behavior and Human Decision Processes, 128*. https://doi.org/10.1016/j.obhdp.2015.02.002

Stadtfeld, C., Takács, K., & Vörös, A. (2020). The emergence and stability of groups in social networks. *Social Networks, 60*, 129–145. https://doi.org/10.1016/j.socnet.2019.10.008

Steffen, E., & Vogel, A. (2021). Concepts of signed graph coloring. *European Journal of Combinatorics, 91*, 103226. https://doi.org/10.1016/j.ejc.2020.103226

Sytch, M., & Tatarynowicz, A. (2013). Friends and foes: The dynamics of dual social structures. *The Academy of Management Journal*. https://doi.org/10.5465/amj.2011.0979

Tang, J., Chang, Y., Aggarwal, C., & Liu, H. (2016). A survey of signed network mining in social media. *ACM Computer Survey, 49*(3). https://doi.org/10.1145/2956185

Valente, T. (2012). Network interventions. *Science, 337*, 49–53. https://doi.org/10.1126/science.1217330

Venkataramani, V., Labianca, G., & Grosser, T. (2013). Positive and negative workplace relationships, social satisfaction, and organizational attachment. *The Journal of Applied Psychology, 99*. https://doi.org/10.1037/a0034090

Wasserman, S., Faust, K., Press, C. U., Granovetter, M., of Cambridge, U., & Iacobucci, D. (1994). *Social network analysis: Methods and applications*. Structural Analysis in the Social Sciences. Cambridge University Press.

Zahedinejad, E., Crawford, D., Adolphs, C., & Oberoi, J. S. (2019). Multiple global community detection in signed graphs. In *Proceedings of the Future Technologies Conference* (pp. 688–707). Springer.

Zheng, X., Zeng, D., & Wang, F. Y. (2015). Social balance in signed networks. *Information Systems Frontiers, 17*(5), 1077–1095. https://doi.org/10.1007/s10796-014-9483-8

Zhou, H., Zeng, D., & Zhang, C. (2009). Finding leaders from opinion networks. In *2009 IEEE International Conference on Intelligence and Security Informatics* (pp. 266–268). IEEE.

Implementation of the Digital Twin of Humans

Adaptive Assistance Systems: Approaches, Benefits, and Risks

Victoria Buchholz and Stefan Kopp

Abstract Digital assistance systems have become ubiquitous in almost all areas of life, supporting us in the execution of tasks at work or in our everyday life. More and more of them are also able to adapt to the needs of the individual user and the environment. Although this development offers new possibilities and provides many benefits, it also poses challenges, as these kinds of systems not only have an impact on the performance of the user for a specific task but also affect psychological factors such as the user's mental workload, self-efficacy, or satisfaction with the system or the task at hand. Moreover, research has shown that inappropriate levels of trust in or reliance on a digital assistance system (i.e., too high or too low) can lead to serious errors and even accidents. To provide effective and acceptable adaptive assistance, technical systems must first keep track of the state of the task, the environment, and the user and then take appropriate actions at the right time in order to improve the performance criteria of interest. The latter involves some form of interaction with the user and may include, e.g., informing, suggesting, intervening, or taking over control. This chapter provides an overview of different adaptive assistance systems, assistance types, and applied strategies, and it discusses their potential importance for the design of adaptive assistance systems. A categorization of different types and systems is presented; they are characterized according to the areas of use, the applied methods and technological implementations, the input data, and the possible effects on the user or the interaction. Concrete examples from our own research on adaptive assistance in monitoring tasks will be discussed.

Keywords Assistance systems · Adaptive systems · Adaptive assistance · Industry · Digital twin

V. Buchholz (✉) · S. Kopp
Social Cognitive Systems Group, Bielefeld University, Inspiration 1, 33615 Bielefeld, Germany
e-mail: vbuchholz@techfak.uni-bielefeld.de

S. Kopp
e-mail: skopp@techfak.uni-bielefeld.de

© The Author(s), under exclusive license to Springer Nature Switzerland AG 2023
I. Gräßler et al. (eds.), *The Digital Twin of Humans*,
https://doi.org/10.1007/978-3-031-26104-6_6

1   Introduction

Humans increasingly receive support from technological devices in almost all parts of their private life and work life. Today, these so-called digital assistance systems (digital ASs) facilitate everyday tasks like navigating from A to B (Lee & Cheng, 2008), support learning (Schodde et al., 2019), assist in manufacturing processes (Eder et al., 2020), provide information, or simply entertain us (Kopp et al., 2005). The use of ASs can be especially helpful and even crucial at times when the safety of the user or of others is at risk. Whether an AS is used as intended or, on the contrary, even disabled depends on the user's acceptance of it. Hence, achieving a high user acceptance is of great importance. Employing an adaptive system can help to reach this goal (Fleming et al., 2019). These types of systems adjust to the individual user and environmental parameters instead of reacting in the same way for everyone at every time (Wandke, 2005).

Depending on the main purpose, the application scenario, and the target group of the AS, the design of an AS can vary greatly from a cute virtual agent to an industrial robot arm to an automated system operating in the background. However, not only the design but also the input and output modalities used for the interaction with the world and the user vary. Some systems only react to voice, others react to touch, and still others require the use of an input device like a computer mouse; some even offer more than one interaction modality. Responses from these systems might be given in the form of sounds like warning tones or even spoken text, written messages, or changes in visual features, such as color changes or changes in facial expressions. Thus, it is not surprising that a substantial body of research focusing on the development of (adaptive) assistance systems in different areas of our lives exists.

Due to the wide range of application fields of these systems, a lack of clarity in the definition and understanding of these systems exists. The German Federal Agency for Occupational Safety and Health, for example, defines adaptive worker assistance systems as "methods, concepts, and computer-aided working systems that support work activities depending on the context and in some cases even autonomously" (original text in German, cf. BAUA – Bundesanstalt für Arbeitsschutz und Arbeitsmedizin (2015, p. 11), translation by this author). This definition points out that the main task of an AS is to support the user but also that it can act autonomously at times and perhaps take over the execution of a task. According to our view, this definition not only applies to worker ASs but also to systems supporting tasks in everyday life. Moreover, this is a very broad definition but it provides an idea of what makes an AS adaptive: The support is context-dependent. To be more specific, an AS can be defined as adaptive when the system itself adjusts to the individual user and the environment.

In contrast, adaptable systems allow each user to adjust the AS according to their own needs; for example, a user can choose their preferred route options in a navigation system. Other forms of ASs also include fixed systems that react in the same way for every user and systems that are customized for specific user groups or

Fig. 1 Driver fatigue detection alert (Miles Continental, 2022)

tasks (Wandke, 2005). All of these types of ASs have advantages and disadvantages and are more or less suited to different contexts. From an application and interaction point of view, adaptive ASs provide an interesting field of research and a promising approach to reaching a high user acceptance (Fleming et al., 2019). In the following, we name a few examples of research on adaptive ASs in different domains, from everyday life to work, to demonstrate the wide range of application fields and the great variety in the appearances of these systems.

A huge amount of research on the design and implementation of driver assistance systems (Bengler et al., 2014) that we use in our everyday lives exists. The developers of such systems are attempting to increase the safety of people inside and outside of the car, reduce the costs of driving and environmental pollution, and increase the mobile efficiency by reducing the energy consumption, the time needed to get from A to B, and the consumption of resources. These systems include, for instance, navigation systems, parking assistants, adaptive cruise control, driver fatigue detection (see Fig. 1), and collision avoidance systems. Some of these ASs are fixed; these systems are often those involving higher risks, like collision avoidance systems. Others can be adapted by the user, who, for example, can select the minimum distance to a leading vehicle in adaptive cruise control, or the systems themselves are adaptive, like fatigue alerts. In this case, the control over the timing of the warning always lies with the system and cannot be adjusted by the driver. Car manufacturers use different sensors and parameters to decide whether the user needs to be warned that he or she might fall asleep soon and should take a break. Miles Continental, for example, states that their system "monitors driver behaviour closely, noting any erratic steering wheel movements, pedal use and any lane deviations" (Miles Continental, 2022).

Another aspect of our everyday life that can be supported by ASs is learning. In this case, the adaptation of the system to the individual needs of the learner can be very beneficial. For example, Schodde et al. (2019) used a small robot to assist

Fig. 2 The virtual assistant displays the user's appointments on a calendar (Yaghoubzadeh et al., 2013)

children in second-language learning, thereby facilitating the process of learning new vocabulary words. Their results demonstrate that the children learned the presented vocabulary words better with the support of the robot and individual adaptations to their current cognitive and affective states and levels of engagement.

Additionally, adaptive ASs exist to support people who are no longer able to do everyday tasks on their own. Yaghoubzadeh et al. (2013), for instance, developed a virtual agent to assist elderly and cognitively impaired people in planning their days and following a daily schedule. In this scenario, a relatively simple but friendly looking virtual agent that stood next to a calendar displayed on a screen was used (see Fig. 2). The agent was able to communicate with the user using speech and adjusted to the speed and understanding of the individual by waiting for feedback before moving on.

Adaptive ASs also play an important role in the work environment, as these systems can help to integrate skilled workers without requiring a human supervisor or specially trained instructor (cf. Besginow et al. (2018)).

In contrast to assistance systems for drivers, ASs in an aviation context already work more autonomously due to a more controlled environment without pedestrians, cyclists, or road bumps. An overview of the systems used in manned and unmanned aircraft is presented by Lim et al. (2018). They described systems that augment information in the cockpit, filter data, offer support in decision-making in time-critical situations, issue warnings when abnormal events occur, plan flights, adapt to the pilot's workload, etc. Additionally, the developed ASs make it possible to extend flight operations, for instance, by making flights in difficult weather conditions possible or by ensuring that pilots can concentrate longer by reducing their mental workload.

Schwarz and Fuchs (2017) developed an adaptive assistance system that considers not only the mental workload of the user but also other psychological factors that influence the user's performance, namely their situation awareness, attention, fatigue, motivation, and emotional state. During the execution of a task, the AS monitors the user's performance. If a performance decline is detected, the system determines possible reasons for this by analyzing the user's states and additional context information. On this basis, the AS should be able to adapt to the individual user and select an appropriate assistance strategy in the future. A simplified version of the system was tested with a few participants for an air surveillance task. The results showed that the reasoning of the AS, which used only the mental workload, passive task-related fatigue, and attention user states, tended to provide a good basis for assistance selection.

Reducing the work load also plays a role in an industrial context. Today, approaches to developing adaptive ASs with augmented reality (AR) exist (Yang & Plewe, 2016). This technology offers more and different possibilities than earlier stationary systems, as the worker can move freely around a work space while support is provided in the form of visualizations that shown in the worker's field of view. Hence, AR technology can be used in assembly tasks and also in other work areas, such as training or maintenance (Eder et al., 2020).

In production processes, time is a critical factor. Shorter innovation cycles, cost pressures (Eder et al., 2020), staff fluctuations, the rising demand for the customization of products, and thus an increasing need for the production of small lot sizes pose a challenge for companies (Oestreich et al., 2021). In order to decrease learning and work times, Oestreich et al. (2021) proposed the introduction of adaptive learning assistance. They developed and investigated the effects of an assistance platform that is able to adapt its instructions to the worker according to the task, the current situation, or the step in the process; it is also able to adapt to the needs of the individual in manual assembly (see Fig. 3).

In the following, we focus on adaptive assistance systems for monitoring tasks, as these tasks can be found in a variety of domains. We start with a short definition of these tasks, take a look at an approach to categorising ASs and assistance types, and describe the general architecture of an AS for these types of tasks, as well as different implementation approaches. Furthermore, we point out the benefits of using ASs in different contexts, possible risks, and the challenges that developers face. This part includes a discussion of the effects of AS usage on different psychological factors. Moreover, we present our work on assistance systems in monitoring tasks. Finally, we summarize the content of this chapter and provide an outlook on future research.

Fig. 3 An assistance scenario in manual assembly (Oestreich et al., 2021)

2 Categorizing Assistance Systems

Given the differences in domains and ASs, it is not surprising that several taxonomies for AS and assistance types exist in the literature. Some researchers focused their categorizations on the level of autonomy (LOA) of the AS. Sheridan's taxonomy of decision and action selection, for example, is widely known (Parasuraman et al., 2000). The proposed levels, however, can only be used for decision-support ASs and are not applicable for tasks in which no decision selection is required. Furthermore, the division into 10 different levels of support might be too detailed for less complex ASs. Therefore, other researchers have used a reduced number of categories to classify ASs. Carsten and Nilsson (2001), for instance, created four different categories of advanced driver assistance systems:

1. Providing information,
2. Providing feedback or warnings,
3. Intervening,
4. Automated driving.

We argue, however, that the last category should not define an AS, as the described system takes over the entire task without giving the user the option to overrule its actions. The first three categories, however, can be applied to a number of ASs.

All in all, it can be said that many of the taxonomies that have been proposed cannot be used to classify systems that are used in dynamic, time-critical tasks that do not focus on making decisions (for an overview, see Wandke (2005)). Thus, Wandke (2005) proposed a taxonomy that is more general and can be applied to different kinds of ASs in different fields of application. This is achieved by distinguishing six

Table 1 The six stages of human action that can be assisted by an AS and their respective assistance types according to Wandke (2005)

Stage	Description	Assistance types
1	Motivation, activation, goal setting	Activation Coach Warning Orientation
2	Perception	Display Amplification Redundancy Presentation
3	Information integration, generating situation awareness	Labelling Interpreter Explanation
4	Decision making, action selection	Supply Filter Adviser Delegation Take-over Informative execution Silent execution
5	Action execution	Power Limit Dosing Shortcut Input
6	Processing feedback of action results	Feedback Critique

stages of human action that can be assisted by a technical system and by considering the different assistance types that can be applied in each stage (see Table 1). These stages correspond to the steps humans take when choosing, executing, and evaluating an action, namely 'motivation, activation, and goal setting' (Why should I act and what is my goal?), 'perception' (What is happening around me? What do I see, feel, hear, etc.?), 'information integration and generating situation awareness' (What do the perceived signals mean? What is the current state of my environment?), 'action execution', and 'processing the feedback of the action results' (What effects or consequences did my action have?). An AS can support just one stage of action or several of them by implementing one or more of the proposed assistance types.

While assistance types are general categories of assistance, how they are implemented in a specific AS depends on the chosen assistance strategy. Thus, the type describes what should be done, while the strategy defines how this should be done and which methods and tools should be used. The assistance type 'filtering information' could be implemented by choosing a specific filtering algorithm. The strategy describes whether such an algorithm is used and if so, which one is chosen. This means that different assistance strategies can be applied in order to implement each of the proposed assistance types. In some cases, the same strategy may even be applied for different types of assistance. In the first stage, for example, an AS may monitor the drivers' state and react to a detection of fatigue or drowsiness by using the strategy of playing a sound and thereby making the driver alert again (activation assistance). Another AS may use warning assistance in a monitoring task when the human operator needs to react to a deviation from the desired behavior of the monitored system. In this case, playing a sound might also be the most effective strategy. In the second stage, signals that might be overlooked can be amplified or presented in a different way by the AS. Some car manufacturers, for example, equip vehicles with a display that presents passed traffic signs to the driver. The information itself is redundant for drivers but it can ensure that a new traffic sign is not overlooked by the driver. The 'presentation' assistance type may also be implemented using other strategies, such as informing the user of the weather conditions, the state of the car, where to find certain things, etc. In the third action stage, appropriate strategies for the assistance types 'labelling', 'interpreter', and 'explanation' include providing labels, translating texts into other languages, and providing help functions or manuals. The first three assistance types in the fourth stage deal with the amount of preprocessing that is performed on the presented information. The AS could simply offer all possibilities, suggest the most important possibilities, or just present a single alternative and advise the user to choose it. The last four types go a step further: In this case, the system might even execute a part of the task for the user. This can be a short, time-constrained part of the task for which the user still has to confirm the execution or is able to overrule the decision of the system, or it could be a part of the task that is running constantly in the background. In contrast to the fourth stage, where the system might take over the execution of certain actions for the user, in the fifth stage, the system supports the action execution of the user. The augmentation of a drivers' brake force is one example of this type of assistance, as well as increasing or decreasing the speed of the mouse cursor in order to help the user hit a target on their computer screen. Assistance types in the last stage aim to provide the user with feedback about or even a critique of their actions. This can be especially helpful in learning environments.

Although this taxonomy does not focus on the autonomy of the user or the AS, it is an interesting aspect that needs to be examined when one is choosing a certain assistance type. This is especially important when one is designing an adaptive AS and aiming for high user acceptance. The relationship between the assistance types and LOAs of the AS, as well as how they influence—positively or negatively—the users' performance and acceptance of the AS, is explained in greater detail in the next section.

3 Benefits and Risks of Assistance Systems

Introducing digital ASs into everyday life or the work environment can have many benefits. When these systems are carefully designed, they can enhance the users' performance, increase their safety on the road or at workplaces, and enable the inclusion of skilled workers with impairments, as demonstrated by studies from different fields of application (cf. Eder et al. (2020), Besginow et al. (2018), and de Visser and Parasuraman (2011)). In this section, we describe a few examples of user studies conducted with adaptive ASs and examine the results. Therefore, it should be noted that this is not a complete overview of existing research on the topic; it describes the potential of adaptive ASs while also considering the risks of introducing and using these ASs.

de Visser and Parasuraman (2011), for instance, investigated the effects of static and adaptive ASs in a supervising task. Their results show an increase in the users' performance for both types of ASs. Moreover, they concluded that adaptive assistance in particular was able to positively influence the users' trust in the AS and their self-confidence, and it was also able to reduce the user's workload during the execution of the task.

A study on adaptive assistance in combination with AR for air traffic controllers was conducted by Gürlük et al. (2018). The researchers concluded that the benefits of the AS are that it increased the situation awareness and reduced the workload of the controllers.

In a study on the benefits of adaptive ASs, Burggräf et al. (2021) asked for the expert opinions of 132 representatives of companies belonging to the manufacturing sector in Germany, Austria, and Switzerland. More than half of these representatives held management positions. Among other aspects, the participants saw high potential in the use of adaptive ASs and believed that these systems could improve flexibility at work and in training processes, as well as reduce workers' cognitive workload.

Another possible advantage of adaptive ASs has already been mentioned: higher user acceptance (Fleming et al., 2019). Furthermore, Fleming et al. (2019) implied that with the introduction of reliable adaptive assistance for car drivers, not only the user acceptance but also the road safety will increase.

Table 2 shows a summary of the presented advantages of using adaptive ASs.

Developing ASs, however, poses multiple challenges as well. While performance optimization or an increase in safety are important design goals in industrial work settings, aviation contexts, or driving, a critical factor that might determine if the system is used as intended in the long term is user acceptance (Fleming et al., 2019). Trösterer et al. (2014) identified an important design principle that can help assistance systems achieve high user acceptance: The system must only provide assistance when it is needed and adapt to the individual needs, capabilities, and preferences of the user. If the AS provides assistance all the time, this might decrease the users' satisfaction with the AS and annoy them after a while. We would go a step further than Trösterer et al. (2014) and argue that the LOA of the AS should also be taken into account when developing these systems. A high LOA means that the AS takes

Table 2 Benefits of the use of adaptive assistance systems

Adaptive ASs improve
Performance (de Visser & Parasuraman, 2011)
Situation awareness (Gürlük et al., 2018)
Safety (Fleming et al., 2019)
Trust (de Visser & Parasuraman, 2011)
Flexibility at work (Burggräf et al., 2021)
Self-confidence (de Visser & Parasuraman, 2011)
Training processes (Oestreich et al., 2021; Burggräf et al., 2021)
User acceptance (Fleming et al., 2019)
Inclusion of workers with impairments (Besginow et al., 2018)
Workload (Gürlük et al., 2018; de Visser & Parasuraman, 2011; Burggräf et al., 2021)

control of the task execution, while a low LOA means that the user is completely in control. In the first case, problems from human-out-of-the-loop performance may arise and lead to serious errors and severe accidents. Especially in monitoring tasks, situation awareness problems like failing to detect a deviation of the behavior of the system from the norm and failing to understand the deviation need to be considered. Furthermore, if the automatic system fails, human operators might not be able to react appropriately and quickly enough to prevent negative outcomes (Kaber & Endsley, 2004). When ASs for pilots were first introduced, problems and severe accidents occurred due to the malfunctions of automated systems, over-reliance on these systems, and detection errors made by the human operators (Wiener & Curry, 1980).

Another term that is often used in this context is complacency. This phenomenon occurs when a system is highly reliable but not perfect. In this case, users might fail to detect abnormalities in the behavior of the system because they feel no need to keep monitoring it (Parasuraman et al., 1993). The risk of accidents increases when complacency occurs in combination with other unfavourable conditions like the fatigue of an operator, heavy traffic, or even bad weather.

The concepts of over-reliance and over-trust and their consequences have also been investigated in advanced driver assistance systems (Inagaki & Itoh, 2010). Apart from the negative effects mentioned previously, both concepts can also lead to an overestimation of the capabilities of the AS when the user does not have sufficient knowledge about it or experience with it. Thus, the authors concluded that it can be useful to equip the AS with multiple layers of assistance strategies in order to prevent over-reliance and over-trust. Hence, the AS should provide assistance only when it is needed and might start by simply warning the user instead of directly taking over a task, thereby slowly increasing its LOA if needed. We will take a closer look at how this can be achieved practically in the next section.

Mahr and Müller (2011) described the possible negative effects of advanced driver assistance systems. They mentioned over-reliance but also considered

Table 3 Risks of the use of (adaptive) assistance systems

Risks
Over-reliance (Wiener & Curry, 1980; Inagaki & Itoh, 2010)
AS is not used (Fleming et al., 2019)
Over-trust (Inagaki & Itoh, 2010; Gürlük et al., 2018)
Complacency (Parasuraman et al., 1993; Gürlük et al., 2018)
Decrease/shift in attention (Mahr & Müller, 2011)
Sudden changes in workload (Mahr & Müller, 2011)

attention decreases or shifts and transition problems. In cases in which the assistance system is working as expected, the users' attention might decrease or they might shift their attention to other tasks. This, in combination with an unexpected failure of the automated system, can lead to a transition problem. In this case, the drivers' mental workload suddenly changes from underload to overload. These possible negative effects can lead to serious errors and accidents. Therefore, it is crucial to consider them when developing an AS.

Table 3 shows a summary of the presented risks and challenges of introducing and using adaptive ASs.

To summarize, the introduction of ASs has risks that need to be kept in mind and in the best case avoided. One way to do this might be to use an adaptive AS and a human- rather than technology-centered design approach for the development of the system. This way, developers can achieve a high user acceptance and counter the possible negative effects of the use of digital systems that support the user in the execution of tasks.

4 Architecture of an Adaptive Assistance System

The basis for the decision to give assistance to the user during the execution of the task and for determining which assistance strategy is appropriate for the given situation is the environment model containing information about environmental parameters, the task, and the user. By knowing the task, e.g., the positions of objects on a screen or work table, and/or the environmental parameters, e.g., the distance from an object to a vehicle, and monitoring them, an AS is able to determine the criticality of the situation and thus how high the risk of failure is. Hence, in order to be able to enhance the user's ability to execute the task, the system needs to monitor, represent, and then analyze their behavior in addition to monitoring the environment. The input data can consist of anything that provides insight into the users' current state. However, these data are also task-specific, as every task offers different data-recording possibilities. In a second step, the AS uses these data to decide whether assistance is needed and if so, which strategy is appropriate for the current situation. In the following, we

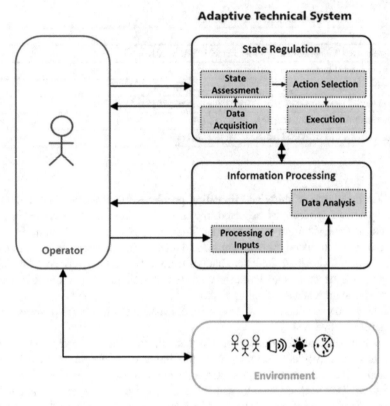

Fig. 4 Simplified model of an adaptive assistance system presented by Schwarz and Fuchs (2017, p. 389)

call this part of the architecture the decision module. Different approaches, some of them task-specific, can be applied here. The last step involves the output. Depending on the output of the decision module, this output can be an assistance strategy or nothing at all. In addition to possibly providing assistance, the AS might also try to communicate why it made a certain decision to the user. The output can then be fed into the input module again to obtain feedback about the effects of the decision that was made.

One example of an architecture for an adaptive AS was proposed by Schwarz and Fuchs (2017) (see Fig. 4). The simplified model shows an 'information processing' component that processes input from the user/operator, the task, and the environment and saves the information in the environment model. These data can then be analyzed and displayed to the operator or used to provide adaptive assistance in the execution of the task. Schwarz and Fuchs (2017) called this component 'state regulation', and it consists of four stages: data acquisition, state assessment, action selection, and execution. State assessment refers to the analysis of the user's current mental or affective state, which provides a basis for the selection and implementation of appropriate assistance strategies.

4.1 Data Acquisition

A variety of possibilities for recording data from the user, the task, and the environment exist. Recording data from the task itself can be very specific; it can range from tracking the positions of objects on a computer screen to tracking the temperatures of machines in a factory. The same applies to environmental parameters, which can range from measuring the distance to the obstacles surrounding a vehicle to looking at a weather forecast. Recording data from the user is a more general task, and solutions can be applied to very different contexts. Töniges et al. (2016), for example, presented five different categories of approaches for gathering data from the user: task-related data, voice data, vital parameters, head movements and facial expressions, and multimodal features. The first approach looks at task-related data, which can consist of performance data like errors or response times and also data recorded from an input device. The second category considers the analysis of the human voice, as this can tell us about the current mental state of the user. However, not all tasks require voice input or even constant speaking. In some environments, the background noise would also make analyzing the audio nearly impossible or at least very difficult. Then, data from other categories need to be examined. The third category of approaches, analyzing vital parameters using physiological sensors, is a very promising approach for different areas of use. Vital parameters can be recorded constantly and directly correlate with human states. Popular methods include eye tracking; measurements of the body temperature, oxygen in the blood, and the galvanic skin response; and electroencephalography. Today, there exist many recording devices that can be integrated into the work station or the work clothes and thus do not interfere with the execution of the task. Eye tracking in particular can be very useful in many areas, as the data can be recorded constantly and unobtrusively, thereby avoiding interference in the execution of the task. The recording devices can be installed on a work station or in a vehicle or can even be worn as glasses. In a previous work, we used this method to acquire information about the user's mental state in order to trigger adaptive assistance (Buchholz & Kopp, 2020). Moreover, eye-tracking information is also used as input for ASs in other domains, such as the aviation (Lim et al., 2018) and automotive industries (Marquart et al., 2015), to infer the operators' mental load. The fourth category represents the analysis of head movements and facial expressions. Head analysis is a promising approach, as human states like boredom, confusion, or physical pain can be detected. In driver assistance systems, head tracking has been applied to detect inattention or fatigue. For many other scenarios, however, suitable head tracking or face recording devices have not yet been developed. The last category of approaches presented by Töniges et al. (2016) consists of the analysis of multimodal features to enhance the robustness of the results. In this case, experiments are needed to clarify which input data are necessary and useful and which are redundant.

All of these approaches enable the monitoring of the user. However, some of the data are very sensitive and might not be legally attainable in some countries and contexts. Thus, the decision concerning which approach is most suitable for a specific adaptive AS depends not only on the goal of the AS, what can be inferred from the different data, and which sensors can be used in the specific task or context, but also on legal and ethical questions.

4.2   Environment Model and Digital Twin

As mentioned previously, the environment model contains information not only about environmental parameters but also about the task and the user. The model can be constructed from the recorded data and additional prior knowledge about the task at hand. In some cases, simple calculations are enough to determine the criticality of the situation from these data; for instance, a system may determine if the collision of a car with an obstacle ahead of it can still be avoided given the distance to the obstacle, the current speed, the possible brake force, and maybe even the condition of the ground that the car is driving on. Creating a digital representation of the user, however, can be more difficult, as this often includes making predictions about the user's cognitive and affective states and behavior. In the following, this digital representation containing all the information about the user that is relevant to the decisions of the adaptive AS is described as the digital twin of the human.

In a previous work, we proposed using the analysis of eye movements (Buchholz & Kopp, 2020) to gain insight into the user's mental state. In an experiment, we were able to demonstrate that eye-tracking parameters, like the number and duration of fixations, and performance data, such as the average reaction time and error rate, from a monitoring task were correlated with the mental workload of the user. This is in line with findings from other researchers: de Greef et al. (2009), for instance, were able to show that certain eye-tracking parameters were correlated with the users' mental workload. They proposed using these findings to trigger adaptive automation.

In addition to detecting the operators' workload, mental states like drowsiness, fatigue, and inattention are of interest to researchers. Different factors can lead to these states. A high workload due to inexperience, having to do more than one task at once, or unusual conditions, for instance, may cause mental fatigue after some amount of time, while a low workload due to repetitive tasks, normal conditions, and experience might lead to drowsiness (Gimeno et al., 2006). Gimeno et al. (2006) provided a good overview of these two concepts, which are called overload and underload, and approaches to measuring them during driving.

Some research has investigated correlations between task-related features, like the movements of a computer mouse, and a user's affective states. The results from studies by Yamauchi and Xiao (2018) and Salmeron-Majadas et al. (2014), for example, suggest that a correlation exists. Like eye-tracking data, task-related features can be obtained unobtrusively. Moreover, no further hardware needs to be integrated. Mostly, however, they are used to increase the validity of recorded data from different sensors (Holmqvist et al., 2011).

In summary, much research has already focused on determining the user's mental and physical states from different input data. It is still a challenge to find the right approach and equipment for a specific task or environment. Additionally, many research results come from laboratory experiments. Therefore, solutions need to be found for factors that might influence measurements in natural environments, such as changing light conditions in eye tracking, noise in speech recognition, or workers moving during the execution of a task in camera tracking.

4.3   Decision Module: Assistance Strategies

For the decision module, two questions need to be answered: *What* types of assistance can and should be applied and which approach or approaches should be used to decide *when* to provide these types of assistance? As we showed in the second section, different types of assistance exist (see Table 1) and various strategies can be applied to implement them. Especially for time-critical tasks, determining when to provide what type of assistance is of great importance. This applies to avoiding collisions during driving, reacting to deviations of autonomous systems from the desired behavior in airplanes, and avoiding machine failures in a factory.

In time-critical tasks, assistance types like coaching, presentation, explanation, or adviser assistance, for example, might not be very useful, as the user does not have enough time to perceive and interpret the presented information or evaluate the advice of the AS. Thus, warning assistance from the first stage of human action might be more suitable for these scenarios. In this case, the LOA of the AS is still very low and the user is completely in command of the task, but the system makes it clear that an immediate action is required from the user. In this case, the user is still responsible for deciding if the warning is appropriate and what kind of action is required. Hence, a sufficient amount of time to react and respond to the critical situation needs to be given. Otherwise, the assistance might be redundant or could distract the user from an important task. A possible strategy for this type of assistance is attention guidance through optical or acoustic cues, as in the assistance types of the second stage. All of these types of assistance are meant to improve the users' performance or turn their attention to important events and information. An assistance type that actively intervenes in the execution of the task is meant to avoid failures like collisions with other vehicles, an aircraft hitting the ground, or the breakdown of a machine in a factory by enhancing the users' actions or taking over completely. This might happen when there is not enough time for the users to react on their own or if the AS believes that a user failure has occurred. In these cases, the LOA of the AS is high, and the users' sense of autonomy and responsibility might decrease. Therefore, we suggest using this type of assistance only when the probability that other types of assistance will lead to a successful outcome is very low. Otherwise, the acceptance of the AS might decrease and the user may stop using the AS completely (Fleming et al., 2019).

Different techniques and approaches, including various machine learning approaches, can be used to determine the appropriate point in time at which to provide assistance, i.e., the stage that should be supported, including various machine learning approaches. Machine learning makes it possible not only to infer the users' mental states or emotions from the given data but also to predict future states and actions. Moreover, these algorithms are able to detect patterns in the data that a human might not be able to see (Wuest et al., 2016). Despite these advantages, according to a recent review on machine learning in manufacturing (Fahle et al., 2020), research on assistance systems in this domain has not mentioned the implementation of these algorithms yet. In contrast, research in other domains has already investigated the suitability of these algorithms extensively.

In order to make sure that assistance is only given when it is really needed—when the user is not able to complete the task by himself or herself—McCall and Trivedi (2006) developed a braking assistance system for cars that considers not only the criticality of the situation, i.e., parameters like the speed of the vehicle and the distance to the next vehicle, but also if the driver intends to brake or not. Therefore, they use two cameras inside the vehicle to monitor the drivers' feet and head. By using a Bayesian framework, their AS was able to determine whether an intervention was needed or not.

Hajek et al. (2013) developed and evaluated workload-adaptive cruise control systems. They used physiological sensors to measure the drivers' heart rate, respiration, and galvanic skin response. In an experiment using a Wizard-of-Oz design (the desired behavior of the technical system is only simulated), a secondary task was used to induce phases of high workload for the driver. During these phases, the AS kept a greater distance between itself and the leading vehicle by gradually reducing the speed until the predefined distance was reached. Thus, the driver had more time to react to events. The questionnaire results showed that the participants preferred the adaptive AS over a static AS that kept the same safety distance throughout the experiment. For the prediction of the workload, sliding windows, i.e., time series that are split into consecutive, overlapping windows, are used as the input for a classification algorithm. First, a forward feature selection algorithm was applied to select the best feature set. These features were then used for a Decision Tree and Naive Bayes classifiers. The decision tree algorithm was able to correctly identify phases of high workload with a relatively high accuracy of 83.7%. In order to apply this approach in real-life scenarios, however, a reference period needs to be recorded beforehand. Thus, the AS would only work reliably after a learning phase and could not be used out of the box. This might affect the user's trust and acceptance of the AS. Furthermore, in work environments, this would mean that workers would either have to work extra hours during the introduction of the AS or would not be able to use it as intended until after the learning phase. Both scenarios could cost a company time and money.

Nevertheless, the presented results are very promising. The use of machine learning, however, can also be challenging, as its success depends on many factors. First of all, an appropriate algorithm needs to be chosen, and training data must be acquired and preprocessed. Some algorithms require a large amount of training data, which can be difficult to obtain in real-life work settings due to data privacy issues (Wuest et al., 2016). Adaptive ASs that adjust their behavior to the individual user in particular might suffer from these challenges.

4.4   Providing Adaptive Assistance and Communicating Decisions

When the decision to provide assistance or change the current assistance type or strategy is made, it has to be executed or presented to the user. Common design principles and previous research on presenting and transmitting information, suggestions, or warnings can be used. In general, different output modalities (visual, auditory, tactile) can be used and also combined to support the user and communicate with him or her.

When an assistance type is not directly visible to the user, for example, when the AS takes over (parts of) the task in the background, the question of whether or not to communicate this information to the user needs to be answered before determining how to communicate this information. Often, there is not enough time to explain the decision of the AS in the moment itself. Thus, one approach would be to inform the user about the functioning and features of the AS ahead of time in order to prevent surprises during the execution of a task. Another approach would be to use visual or acoustic cues to inform the user that the AS has intervened and to explain these cues before the AS is used.

5   Example: Adaptive Assistance in Monitoring Tasks

Due to a change in the role of the worker during the Fourth Industrial Revolution, so-called monitoring tasks have become more and more common in many different domains; the controllers of factory plants, drivers of (semi-)autonomous vehicles, and operators of air or train traffic all perform monitoring tasks. Instead of having workers execute a task themselves, nowadays, an automated system executes the task for the worker, leaving the worker to monitor the system and to react to any deviation from a desired behavior. Especially in an industrial context, this development has not been taken into account in greater detail in recent research. Instead, many studies focus on the development of assistance systems with virtual or augmented reality or mobile solutions (Yang & Plewe, 2016; Gorecky et al., 2014). Thus, in our current research, we focus on the implementation and investigation of an adaptive assistance system for monitoring tasks.

This type of task requires the user to monitor an automated system and to react to abnormalities. This task is dynamic and can be time-critical. Thus, it can be found in various domains and tasks, such as driving a car, flying a plane, or supervising a number of machines in a factory. However, the same type of task can look very different in different contexts and domains. Reacting to events while driving a car and reacting to events in a factory are most likely not directly comparable. Hence, different assistance types and strategies need to be applied to support the user in an optimal way.

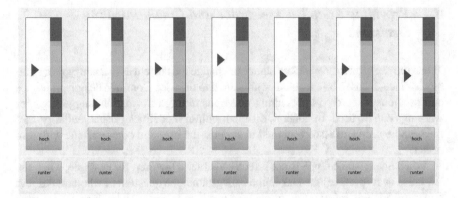

Fig. 5 The monitoring task. The buttons are labeled *hoch* and *runter* ('up' and 'down' in English)

By considering a simulated task in a virtual environment that requires the user to monitor seven temperature gauges and to react to deviations from the optimal temperature by pressing the correct button (see Fig. 5), we are able to test different assistance strategies and examine the effects on the user. In the first experiment, we demonstrated that it is possible to infer a user's level of mental workload during the execution of the task from input data (Buchholz & Kopp, 2020). In this case, the digital twin contained performance and eye-tracking parameters.

For this monitoring task, we are developing an adaptive assistance system that incorporates an environment model to keep track of the user's behavior, the events that are occurring, and the positions of buttons on the screen. The user's behavior is thereby analyzed using the movements of the mouse, the average time the user takes to respond to events, and the number of errors made, as well as the user's eye movements. Using these input data, the situation difficulty and the user's ability to handle the current situation are determined. The AS then has to decide whether it needs to interfere and if so, which assistance strategy should be applied.

Based on similar approaches for driver assistance systems, we propose a division of the task into four phases (see Fig. 6). As the task is time-critical, we recommend applying the warning and intervening assistance types with different assistance strategies. Intervening types involve the situation in which the AS acts more autonomously, as it does in the action execution stage and when it completely takes over the execution of the task. This approach and the division into several task phases is similar to the design ideas for emergency brake assist systems in modern cars. Depending on the manufacturer, the number of phases, as well as the applied strategies, vary. Typically, however, the AS warns the driver of a probable collision before intervening in the driving process (Dávideková et al., 2017). In a subsequent phase, the AS can decide to support the driver by increasing the applied braking force or even by taking control and stopping the car autonomously if the driver does not react in time (Bengler et al., 2014).

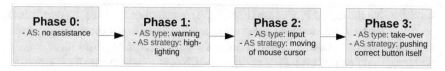

Fig. 6 The four phases of adaptive assistance, with corresponding strategies for supporting the user in the execution of the task

In contrast to emergency brake assist systems, where the phases changes over the course of one single event (a possible collision), however, these four phases will appear over a number of events (deviation of the system from its desired behavior) in our monitoring task, as changing the type of assistance over a few seconds (the length of an event) could be very irritating and confusing for the user.

In the first phase (Phase 0), no assistance is needed, as the monitored system is running smoothly and the worker is able to respond to events in time without making errors. When an event occurs, i.e., the automated system deviates from the desired behavior, an action by the worker is required and thus, assistance might be needed. The AS will go from Phase 0 to Phase 1 if the analysis of the data gathered in the digital twin suggests that the user needs support. In this second phase, the worker is still able to respond to the deviation correctly and in time without any support, theoretically. However, as time is a critical factor in the task and failures should be avoided, the AS can decide to issue a warning to improve the user's performance. This decision could depend on the current average response time or the perceived mental workload of the user. Should these parameters decrease (increase in the workload), the AS might decide that a warning is not enough because data from the previous events demonstrate that there is not enough time for the user to process a warning and then act accordingly. Thus, this marks the transition into the third phase (Phase 2). In this phase, the AS can intervene by supporting the user, e.g., by pulling the mouse to the target. This way, it is more likely that a failure will be avoided, and the user is still in charge of the action (the clicking of the correct button). When the user is not able to appropriately respond in time even with the support of the AS over a window of events, the AS moves to the last phase (Phase 3). It will simply take over the task for the user to avoid failures. If this leads to an improvement of the user parameters, such as the mental workload, the AS will return to one of the previous phases.

By performing this division into different phases instead of letting the AS simply take over the task for the user frequently, we are trying to prevent a loss of the user's sense of responsibility and autonomy during the execution of the task, as the user is still in charge and will be held responsible for any failures. Whether or not this is actually what occurs needs to be investigated in future studies.

In order for the AS to provide the best support for the worker, it is crucial to determine the transitions between the different phases reliably, based on the users' behaviour and the task situation. We have started to test different machine learning algorithms as a first approach, but we have also considered straightforward, heuristic solutions like the use of Fitts' Law (Fitts, 1954) to calculate the time the user would need to reach the right button with the mouse, as the available data complicate the use of some of the common machine learning algorithms for time series classification problems in dynamic tasks. First of all, our dataset is relatively small compared to the datasets commonly used in machine learning (one data point corresponds to one time series here, i.e., all data points belong to one task event). Second, we are dealing with imbalanced data—when the time series is divided into two classes, 'success' and 'failure', the second class consists of fewer time series than the first one. Furthermore, the overlapping of events and the quick changes in the task complicate the detection of distinct patterns in the data of the two classes.

6 Conclusion and Future Work

Today, assistance systems can (and will) be found in all parts of our lives: at home, while traveling, and at work. Adaptive systems will emerge more and more in different domains. On the one hand, these systems have many advantages, as they are able to increase our safety, improve our performance, decrease our mental workload, and remove barriers to the inclusion of workers with impairments. On the other hand, however, they can also have negative effects; they can cause inattention, over-reliance, over-trust, and complacency, which can result in a loss of competence and serious accidents. Furthermore, without the acceptance of the AS by the user, the system will not be used in the long term. Following a human-centered approach, an adaptive AS that only provides support when it is really needed and that can adjust to the individual user should be developed. Thereby, it is likely that workers' sense of autonomy and responsibility can be retained and that the user acceptance of the AS can be increased. Different assistance types and strategies can be applied to reach this goal. In this chapter, we gave an overview of the different assistance types and implementation technologies that can be used for the design and implementation of these assistance types in an AS. Moreover, we presented an example of an adaptive assistance system for a monitoring task, and we described a division of the assistance provided by the system into different phases that call for different assistance types and strategies.

Often, developers and companies look at the benefits of introducing adaptive assistance systems into work environments but lose sight of the possible negative effects and serious consequences that the use of these systems can have. Therefore, it is important to consider all possible effects of a new assistance system so that everybody involved can benefit from its introduction into an environment.

For the construction of the digital twin of the human and the determination of an appropriate time for the assistance system to act, machine learning approaches are very popular and show promising results. However, one has to keep in mind that these approaches also have challenges and that their suitability might depend on the available data.

Acknowledgements Victoria Buchholz and Stefan Kopp are members of the research program 'Design of Flexible Work Environments—Human-Centric Use of Cyber-Physical Systems in Industry 4.0', which is supported by the North Rhine-Westphalian funding scheme 'Forschungskolleg'.

References

BAUA – Bundesanstalt für Arbeitsschutz und Arbeitsmedizin (2015) Forschungsentwicklungsprogramm 2014–2017. https://www.baua.de/DE/Angebote/Publikationen/Intern/I28.html. Retrieved February 8, 2022.

Bengler, K., Dietmayer, K., Farber, B., Maurer, M., Stiller, C., & Winner, H. (2014). Three decades of driver assistance systems: Review and future perspectives. *IEEE Intelligent Transportation Systems Magazine, 6*(4), 6–22.

Besginow, A., Büttner, S., & Röcker, C. (2018). Intelligent adaptive assistance systems in an industrial context–overview of use cases and features. Mensch und Computer 2018-Workshopband.

Buchholz, V., & Kopp, S. (2020). Towards an adaptive assistance system for monitoring tasks: Assessing mental workload using eye-tracking and performance measures. In *2020 IEEE International Conference on Human-Machine Systems (ICHMS)* (pp. 1–6). IEEE.

Burggräf, P., Dannapfel, M., Adlon, T., & Föhlisch, N. (2021). Adaptive assembly systems for enabling agile assembly – empirical analysis focusing on cognitive worker assistance. *Procedia CIRP, 97*, 319–324. https://doi.org/10.1016/j.procir.2020.05.244. 8th CIRP Conference of Assembly Technology and Systems.

Carsten, O. M. J., & Nilsson, L. (2001). Safety assessment of driver assistance systems. *European Journal of Transport and Infrastructure Research, 1*(3), 225–243.

Dávideková, M., et al. (2017). Nice, Berlin, London-If every car had autonomous emergency braking system for forward collisions avoidance. *Procedia Computer Science, 110*, 386–393.

Eder, M., Hulla, M., Mast, F., & Ramsauer, C. (2020). On the application of augmented reality in a learning factory working environment. *Procedia Manufacturing, 45*, 7–12.

Fahle, S., Prinz, C., & Kuhlenkötter, B. (2020). Systematic review on machine learning (ML) methods for manufacturing processes - identifying artificial intelligence (AI) methods for field application. *Procedia CIRP, 93*, 413–418. https://doi.org/10.1016/j.procir.2020.04.109

Fitts, P. M. (1954). The information capacity of the human motor system in controlling the amplitude of movement. *Journal of Experimental Psychology, 47*(6), 381.

Fleming, J. M., Allison, C. K., Yan, X., Lot, R., & Stanton, N. A. (2019). Adaptive driver modelling in ADAS to improve user acceptance: A study using naturalistic data. *Safety Science, 119*, 76–83.

Gimeno, P. T., Cerezuela, G. P., & Montañés, M. C. (2006). On the concept and measurement of driver drowsiness, fatigue and inattention: Implications for countermeasures. *International Journal of Vehicle Design, 42*(1–2), 67–86.

Gorecky, D., Schmitt, M., Loskyll, M., & Zühlke, D. (2014). Human-machine-interaction in the Industry 4.0 era. In *2014 12th IEEE International Conference on Industrial Informatics (INDIN)* (pp. 289–294).

de Greef, T., Lafeber, H., van Oostendorp, H., & Lindenberg, J. (2009). Eye movement as indicators of mental workload to trigger adaptive automation. In *International Conference on Foundations of Augmented Cognition* (pp. 219–228). Springer.

Gürlük, H., Gluchshenko, O., Finke, M., Christoffels, L., & Tyburzy, L. (2018). Assessment of risks and benefits of context-adaptive augmented reality for aerodrome control towers. In *2018 IEEE/AIAA 37th Digital Avionics Systems Conference (DASC)* (pp. 1–10). IEEE.

Hajek, W., Gaponova, I., Fleischer, K., & Krems, J. (2013). Workload-adaptive cruise control – a new generation of advanced driver assistance systems. *Transportation Research Part F: Traffic Psychology and Behaviour, 20*, 108–120. https://doi.org/10.1016/j.trf.2013.06.001

Holmqvist, K., Nyström, M., Andersson, R., Dewhurst, R., Halszka, J., & van de Weijer, J. (2011). *Eye tracking: A comprehensive guide to methods and measures*. Oxford University Press.

Inagaki, T., & Itoh, M. (2010). Theoretical framework for analysis and evaluation of human's over-trust in and over-reliance on advanced driver assistance systems. *Proc HUMANIST*.

Kaber, D. B., & Endsley, M. R. (2004). The effects of level of automation and adaptive automation on human performance, situation awareness and workload in a dynamic control task. *Theoretical Issues in Ergonomics Science, 5*(2), 113–153. https://doi.org/10.1080/1463922021000054335

Kopp, S., Gesellensetter, L., Krämer, N. C., & Wachsmuth, I. (2005). A conversational agent as museum guide - design and evaluation of a real-world application. In T. Panayiotopoulos, J. Gratch, R. Aylett, D. Ballin, P. Olivier, & T. Rist (Eds.), *Intelligent virtual agents* (pp. 329–343). Berlin Heidelberg: Springer.

Lee, W. C., & Cheng, B. W. (2008). Effects of using a portable navigation system and paper map in real driving. *Accident Analysis & Prevention, 40*(1), 303–308. https://doi.org/10.1016/j.aap.2007.06.010

Lim, Y., Gardi, A., Sabatini, R., Ramasamy, S., Kistan, T., Ezer, N., Vince, J., & Bolia, R. (2018). Avionics human-machine interfaces and interactions for manned and unmanned aircraft. *Progress in Aerospace Sciences, 102*, 1–46.

Mahr, A., & Müller, C. (2011). A schema of possible negative effects of advanced driver assistant systems. In *Proceedings of the Sixth International Driving Symposium on Human Factors in Driver Assessment, Training and Vehicle Design*. University of Iowa.

Marquart, G., Cabrall, C., & de Winter, J. (2015). Review of eye-related measures of drivers' mental workload. *Procedia Manufacturing, 3*, 2854–2861.

McCall, J. C., & Trivedi, M. M. (2006). Human behavior based predictive brake assistance. In *IEEE Intelligent Vehicles Symposium, Proceedings* (pp. 8–12).

Miles Continental. (2022). Driver fatigue detection. https://www.milescontinental.co.nz/news/features/driver-fatigue-detection/. Retrieved February 28, 2022.

Oestreich, H., da Silva Bröker, Y., & Wrede, S. (2021). An adaptive workflow architecture for digital assistance systems. In *The 14th Pervasive Technologies Related to Assistive Environments Conference* (pp. 177–184).

Parasuraman, R., Molloy, R., & Singh, I. L. (1993). Performance consequences of automation-induced 'complacency'. *The International Journal of Aviation Psychology, 3*(1), 1–23. https://doi.org/10.1207/s15327108ijap0301_1

Parasuraman, R., Sheridan, T. B., & Wickens, C. D. (2000). A model for types and levels of human interaction with automation. *IEEE Transactions on Systems, Man, and Cybernetics Part A: Systems and Humans, 30*(3), 286–297. https://doi.org/10.1109/3468.844354

Salmeron-Majadas, S., Santos, O. C., & Boticario, J. G. (2014). An evaluation of mouse and keyboard interaction indicators towards non-intrusive and low cost affective modeling in an educational context. *Procedia Computer Science, 35*, 691–700.

Schodde, T., Hoffmann, L., Stange, S., & Kopp, S. (2019). Adapt, explain, engage–A study on how social robots can scaffold second-language learning of children. *ACM Transactions on Human-Robot Interaction (THRI), 9*(1), 1–27.

Schwarz, J., & Fuchs, S. (2017). Multidimensional real-time assessment of user state and performance to trigger dynamic system adaptation. In D. D. Schmorrow & C. M. Fidopiastis (Eds.), *Augmented cognition* (pp. 383–398). Neurocognition and machine learning: Springer International Publishing, Cham.

Töniges, T., Ötting, S. K., Wrede, B., Maier, G. W., & Sagerer, G. (2016). An emerging decision authority: Adaptive cyber-physical system design for fair human-machine interaction and decision processes. In *Cyber-physical systems: Foundations, principles and applications*. Elsevier. https://doi.org/10.1016/B978-0-12-803801-7.00026-2

Trösterer, S., Wurhofer, D., Rödel, C., & Tscheligi, M. (2014). Using a parking assist system over time: Insights on acceptance and experiences. In *Proceedings of the 6th International Conference on Automotive User Interfaces and Interactive Vehicular Applications* (pp. 1–8).

de Visser, E., & Parasuraman, R. (2011). Adaptive aiding of human-robot teaming: Effects of imperfect automation on performance, trust, and workload. *Journal of Cognitive Engineering and Decision Making, 5*(2), 209–231.

Wandke, H. (2005). Assistance in human–machine interaction: A conceptual framework and a proposal for a taxonomy. *Theoretical Issues in Ergonomics Science, 6*(2), 129–155. https://doi.org/10.1080/1463922042000295669

Wiener, E. L., & Curry, R. E. (1980). Flight-deck automation: Promises and problems. *Ergonomics, 23*(10), 995–1011. https://doi.org/10.1080/00140138008924809

Wuest, T., Weimer, D., Irgens, C., & Thoben, K. D. (2016). Machine learning in manufacturing: Advantages, challenges, and applications. *Production & Manufacturing Research, 4*(1), 23–45. https://doi.org/10.1080/21693277.2016.119251

Yaghoubzadeh, R., Kramer, M., Pitsch, K., & Kopp, S. (2013). Virtual agents as daily assistants for elderly or cognitively impaired people. In *International Workshop on Intelligent Virtual Agents* (pp. 79–91). Springer.

Yamauchi, T., & Xiao, K. (2018). Reading emotion from mouse cursor motions: Affective computing approach. *Cognitive Science, 42*(3), 771–819.

Yang, X., & Plewe, D. A. (2016). Assistance systems in manufacturing: A systematic review. In C. Schlick & S. Trzcieliński (Eds.), *Advances in ergonomics of manufacturing: Managing the enterprise of the future* (pp. 279–289). Cham: Springer International Publishing.

Work Autonomy and Adaptive Digital Assistance in Flexible Working Environments

Elisa Gensler, Hendrik Oestreich, Anja-Kristin Abendroth, Sebastian Wrede, and Britta Wrede

Abstract Digital assistance has increasingly been implemented to cognitively support employees and enhance efficiency in different working environments. However, implementations of current assistance systems are often too rigid to provide personalised support, which can restrict employees' work autonomy. Therefore, in this chapter we discuss opportunities to enhance human work autonomy in collaboration with digital assistance systems by facilitating adaptivity within these systems. First, we discuss the concept of autonomy from both a sociological and a technical perspective. Second, we develop a theoretical conception of the interaction between human and system autonomy. We identify three theoretical scenarios and provide examples of autonomy distributions in which increasing technical autonomy (a) restricts human autonomy, (b) does not affect human autonomy, or (c) enhances human autonomy. Based on this conceptualisation, we suggest that adaptive digital assistance promotes flexible distributions of autonomy levels between human beings and systems, which better addresses and enhances human work autonomy. Proceeding from research findings, our previous work, and theoretical assumptions, we present a guideline with concrete design recommendations for adaptive digital assistance systems that focus

E. Gensler (✉) · A.-K. Abendroth
Faculty of Sociology, Bielefeld University, Universitätsstraße 25, 33615 Bielefeld, Germany
e-mail: elisa.gensler@uni-bielefeld.de

A.-K. Abendroth
e-mail: anja.abendroth@uni-bielefeld.de

H. Oestreich · S. Wrede
CoR-Lab, Bielefeld University, Universitätsstraße 25, 33615 Bielefeld, Germany
e-mail: hoestreich@cor-lab.de

S. Wrede
e-mail: swrede@cor-lab.de

B. Wrede
Software Engineering for Cognitive Robots and Cognitive Systems, Faculty 3-Mathematics and Computer Science, University of Bremen, Bibliothekstraße 5, 28359 Bremen, Germany
e-mail: bwrede@cor-lab.de

© The Author(s), under exclusive license to Springer Nature Switzerland AG 2023   137
I. Gräßler et al. (eds.), *The Digital Twin of Humans*,
https://doi.org/10.1007/978-3-031-26104-6_7

on increasing human work autonomy. In this context, we also discuss under what conditions the digital twin of a human can contribute to adaptivity to provide specific support for employees and better interaction opportunities with digital assistance.

Keywords Human autonomy · Work autonomy · System autonomy · Technical autonomy · Shared autonomy · Digital assistance · Adaptivity · Adaptability

1 Introduction

Digital assistance systems are becoming more relevant in both private and professional life. Assistive tools and applications find their way into our private lives via smart home devices or smartphones that have, for instance, digital voice assistants. Digital assistants help to plan users' days (e.g., by accessing weather forecasts, appointments, and to-do lists), inform users about the news, entertain users by telling jokes or playing music, and even guide users in cooking meals. In work contexts, assistance systems support employees by assigning and explaining tasks, recommending decisions, (partially) executing tasks, or automatically storing data and information about work processes. Even though assistance systems in work environments are used to cognitively support workers (Bläsing & Bornewasser, 2021), help employees to assure product quality, and make work processes more productive, they often entail an increase in workers' dependency on even more digital support (Strauch, 2018). As digital assistance systems become more autonomous, they are capable of supporting or even taking over single (complex) tasks. Due to their many functions, digital assistance systems make it possible to change work practices, schedules, and also the criteria for assessing work objectives and their results, which can affect employees' decision-making autonomy. However, work autonomy is an essential job resource, as it reduces the risk of job stress (Kalleberg et al., 2009) and promotes work motivation, as well as job satisfaction (Muecke & Iseke, 2019; Wheatley, 2017). When employees' autonomy is clearly restricted, they may lose track of work situations and thus have difficulties in recognising when they must intervene. As a result, employees may lose the skills and knowledge that they need to control work processes (Selke & Biniok, 2015). Moreover, only humans are able to evaluate whether and to what extent a process or an outcome is meaningful in terms of its content. In addition, autonomy encourages employees to remain alert for possible errors and to optimise processes. This is why the implementation of digital assistance should ensure employees' autonomy, allowing employees to recognise and avoid the unintended consequences of digitally supported work processes.

We suggest that one essential cause of these unintended effects is the often insufficient capability of digital assistance systems to adapt to changing circumstances and the varying skills and needs of human (co-)workers. Therefore, we discuss whether and to what extent system adaptivity better addresses and enhances human work autonomy.

This chapter is structured as follows: First, by defining and contrasting the concepts of human work autonomy and system autonomy in Sect. 2, we expose central similarities and differences. This section considers possible challenges and conflicts in combining human work autonomy and system autonomy. Section 3 specifies the essential aspects of digital assistance and explains its role in the interaction between human and system autonomy. In Sect. 4, by deriving nine possible scenarios, we further elaborate how human work autonomy and system autonomy relate to each other. These scenarios also indicate the applications and technical possibilities of digital assistance and its possible effects, especially for work autonomy. Section 5 illustrates adaptive digital assistance and suggests how adaptive features can be used to address current implementations of too-static digital assistance, which undermines work autonomy. To initiate a discussion and devote attention to the topic of human-centred adaptive assistance, in Sect. 6, we present a guideline for developing adaptive digital assistance systems that specifically consider and thus can enhance human work autonomy. Whether and under what conditions the concept of the digital twin of a human helps to achieve certain levels of adaptivity that promote flexible balances between human work autonomy and system autonomy will be discussed in Sect. 7. Section 8 concludes the chapter and discusses the prospects of relevant research on the topic of adaptive digital assistance that considers human work autonomy.

2   Social Science and Technical Perspectives on Autonomy

Autonomy (from the ancient Greek 'autonomos': autos = self; nomos = law) literally means self-legislation, which can be defined as the capability to make informed decisions using free will that follow self-established rules and regulations. Thus, members of autonomous societies and organisations negotiate, establish, acquire, and set boundaries for their own rules of action (Durkheim, 1984). Transferring this concept of autonomy to a work environment in which humans interact with technical systems, we can distinguish two forms of autonomy: employees' work autonomy and system autonomy.

Human work autonomy refers to employees' decision latitude and the controllability of their own work environment. Employees with high levels of work autonomy can consciously influence work procedures in a more flexible and direct manner to make their own decisions regarding work methods, schedules, and work objectives (Breaugh, 1985). Granting employees work autonomy helps them to cope with work demands and flexibly adapt to unexpected situations by applying their individual experiences and skills (Green, 2008; Parker & Grote, 2020; Pongratz & Voß, 2003). Furthermore, work autonomy is positively associated with job satisfaction (Kalleberg & Vaisey, 2005), work motivation (Muecke & Iseke, 2019), productivity (Green, 2008), innovation (de Spiegelaere et al., 2014), and skill development (Kalleberg et al., 2009). Therefore, it functions as a job resource for employees (Demerouti et al., 2001) that helps to reduce the health-related risks of job strain, especially when their workloads and mental or physical efforts are high

(Demerouti & Nachreiner, 2019; Karasek, 1979). However, research shows that granting employees too much autonomy can also result in high job stress or low performance rates (Kubicek et al., 2014; Langfred & Moye, 2004). For example, Kubicek et al. (2014) showed that too much autonomy can overwhelm employees if they have too much responsibility, and this further deteriorates their work motivation and well-being. This seems to be more likely to occur when employees' self-efficacy is low or the demands on their cognitive attention are high (Langfred & Moye, 2004; Muecke & Iseke, 2019; also cf. Kubicek et al., 2017). These findings suggest that depending on the work situation and personal characteristics, the optimal level of human work autonomy varies (Kubicek et al., 2017; Muecke & Iseke, 2019).

Looking at the topic of autonomy from a technical perspective, we start by defining the characteristics of so-called autonomous systems. An autonomous system must be capable of organising its own behaviour, which includes executing actions, planning ahead, and selecting goals. This goes beyond executing automated procedures and requires interaction with a dynamic environment and therefore finding satisficing (satisfying + sufficient) solutions for problems instead of optimal solutions (Schilling et al., 2016). Alaieri and Vellino (2016) emphasise the ability of autonomous systems to operate without external intervention. Similarly, O'Neill et al. (2020) emphasised self-government and self-directed behaviour (agency), which autonomous agents must fulfil. Furthermore, in situations in which cooperation with humans is necessary, the autonomous agent must take on its own role and work interdependently with human team members to achieve a shared objective (O'Neill et al., 2020). Rosen et al. (2015) mentioned skill-based approaches to modelling capabilities in cyber-physical (production) systems (CP[P]Ss); skills can be used as an indicator of autonomy since the encapsulation of capabilities in skills is a key element of being able to modify the course of action dynamically. Ansari et al. (2018) claimed that the autonomy degree '[...] reflects the cognitive and mechanical capacity of CPPS to create and control the execution of its own strategies and plans as well as to make actions on tasks or detected problems.'

One essential purpose of developing and implementing autonomous systems is to create more flexible systems that go beyond traditional automated systems (Endsley, 2017). While the latter might work very efficiently, the disadvantage of these systems is that they only perform well for very specific tasks with predefined limits and constraints. According to Beer et al. (2014), system autonomy has two main goals: On the one hand, the objective is to let a system perform a desired task well, making human intervention unnecessary. On the other hand, the objective is for the system to be proactively social, meaning that although the autonomous system can run without much human intervention, the system should be capable of flexibly and naturally interacting with its human counterpart. This is especially useful, for example, in situations in which human-machine interaction is indispensable, especially when cooperation is needed.

These definitions show that although both concepts (i.e., human and system autonomy) refer to the definition of autonomy provided above, they have been further developed and have become specific concepts adapted to their applications. In the following (see Table 1), we specify characteristics and point out the similarities and

Table 1 Comparison of human work autonomy and system autonomy

Category	Human work autonomy	System autonomy
Dimensions	• Method • Schedule • Criteria	• Normative control: intentions • Strategic control: plans • Operative control: selection of means
Measurement/assessment	• Subjective perception • Objective assessment by supervisor or job experts	• Definitions with different numbers of levels exist (no automation to full automation) • Criteria for each level define the autonomy degree of the system
Central prerequisites	Opportunities to apply and maintain knowledge and skills to ensure • transparency of work processes • predictability of work demands • controllability	• Ability to perform under uncertainties and in unknown situations and environments • Performance for extended periods of time with limited communication and without external intervention • Knowledge and skills • Situation awareness • Up-to-date user model • Connectivity and interaction
Negotiation	Within the limits of the job in social exchange with employers represented by • managers • supervisors • colleagues • technical systems	Co-construction with • employees (or even their digital twins) • other autonomous agents or rule-based with • other systems

differences between work autonomy and system autonomy based on four categories: dimensions, measurement/assessment, prerequisites, and negotiation. This comparison aims to make clear which aspects involve challenges and which involve opportunities to strengthen human work autonomy in work environments where system autonomy is increasing. It is particularly important to promote human work autonomy in digitally supported work processes, as autonomously acting employees are more alert to errors and can contribute more to improvements and innovation.

2.1 Dimensions of Human Work Autonomy and System Autonomy

Human work autonomy and system autonomy imply different dimensions, which can be combined to define comparable concepts of autonomy. Scholars in the human and social sciences have discussed various dimensions of human work autonomy. Some distinguish between individual autonomy and team autonomy (Kalleberg et al., 2009; Langfred, 2005). Referring to employees' individual autonomy, other scholars distinguish between autonomy as the opportunity to make decisions on work processes and autonomy as the freedom to decide when and where to work (Parker & Grote, 2020; de Spiegelaere et al., 2016). In this chapter, we focus on employees' individual decision-making autonomy, which is especially relevant when studying how employees cooperate with digital assistance systems that also support and make decisions about work processes. Within the concept of human work autonomy, Breaugh (1985, 1999) and similarly Morgeson and Humphrey (2006) considered three individual autonomy dimensions: method autonomy, schedule autonomy, and criteria autonomy.[1] Method autonomy is the extent to which employees can control the procedures, as well as the means or working tools, that they use to perform tasks. Schedule autonomy describes the extent to which employees can decide independently how to schedule their working tasks. This involves the pace, rhythm, and order in which employees perform tasks. Criteria autonomy refers to the extent to which employees determine their work objectives and set the criteria for job performance evaluation (Breaugh, 1985; Morgeson & Humphrey, 2006). This includes the extent to which employees can assess the outcome (i.e., the work quality) of their performance on their own and their ability to influence work objectives.

It is vital to distinguish between the three dimensions of work autonomy depending on the application since each dimension affects several aspects of job quality and work performance (Muecke & Iseke, 2019; de Spiegelaere et al., 2016). The reverse is also true: Work environments do not influence all dimensions to the same degree. For instance, employees working at an assembly line might experience little to no work autonomy regarding how to schedule their work or which methods to apply due to the interaction with the technical system. Still, the job can allow these employees discretion within specified criteria of quality to decide by themselves when and if a product is of good quality, if it needs to be modified, or if it needs to be discarded. Thus, having autonomy in one dimension does not depend on having autonomy in another dimension. Considering the relationship of autonomy to employees' job quality, the three dimensions can even be placed in a flat hierarchical order: Criteria autonomy enhances job satisfaction and work motivation more than method and schedule autonomy (Humphrey et al., 2007; Muecke & Iseke, 2019). Furthermore, schedule, method, and criteria autonomy are particularly relevant dimensions in

---

[1] Some scholars also refer to criteria autonomy using the term 'decision-making autonomy' (Morgeson & Humphrey, 2006; Muecke & Iseke, 2019). We stick to the term 'criteria autonomy' for a clear distinction, as all the described autonomy dimensions involve making decisions about various facets of work.

socio-technical systems in which employees share functions, meaning that their levels of decision-making autonomy can interact with the technical system's autonomy.

While human autonomy is often discussed with respect to situational conditions that enable humans to 'live out' their autonomy and make decisions independently, the general ability to act in an autonomous way (e.g., moving and navigating in unknown environments, being able to freely communicate, etc.) is often part of the autonomy definitions of technical systems. This is also reflected in the ways that autonomy is defined: The definition of the work autonomy of humans distinguishes between different aspects of autonomy (criteria, schedule, method), while system autonomy definitions like that of Gransche et al. (2014) emphasise hierarchies of autonomy levels that build on one another. This can be related to the architectural patterns of autonomous systems in which, for example, reactive and deliberative layers control different parts and aspects of a system (reactive = control and access to sensors and actuators, deliberative = higher-level planning and reasoning) (Murphy, 2000, p. 257). While humans and animals become autonomous agents in their environment after a nursery period (Bibel, 2010), for technical systems, these fundamentals of autonomy have to be implemented by developers. In specific implementations, control aspects might be realised as so-called skills (Rosen et al., 2015) and would be located in the lowest level of the hierarchy (operative control). The middle level describes strategic decisions and control. This includes planning algorithms and mechanisms for choosing a solution from among alternatives. The top level describes acting within norms and rules and the awareness that these norms and rules exist and must be respected. While systems might be able to model and learn these rules and can even evaluate their own behaviour using these rules, systems are not capable of interpreting the deeper meaning of norms (normative control). Thus, Gransche et al. (2014) concluded that the highest level of autonomy cannot be ascribed to systems alone.

For the sake of completeness, it should be mentioned that other distinctions for autonomy dimensions also exist, like that of Gottschalk-Mazouz (2008), which lists functionalities like energetic autarchy, mobility, automation, environmental independence, adaptivity, learning, innovation, and unpredictability. However, as these dimensions mostly focus on interactions with the physical rather than the social environment, we do not take them into account in our discussion.

Comparing the central dimensions of human work autonomy and system autonomy shows that different technical terms are used in the two disciplines, even if the concepts denote similar concepts in some ways. This can be challenging for interdisciplinary discussions of autonomy and requires much communication and explanation.

2.2 Assessment of Human Work Autonomy and System Autonomy

To enhance human work autonomy in flexible work environments in which technical systems also work autonomously, it is vital to be able to assess their respective autonomy levels. Employees' level of autonomy can be assessed using objective as well as subjective measurements (Hacker & Sachse, 2014). Some studies assess employees' work autonomy by having job experts observe employees performing tasks (Schweden et al., 2019) or by asking employees' supervisors to assess the autonomy granted to employees (Choi et al., 2008). These types of assessments are supposed to be more objective; nevertheless, they can differ from personal perceptions of autonomy (Hacker & Sachse, 2014; Schweden et al., 2019). Most studies apply more subjective measurements; for example, employees are asked to rate their agreement with statements about perceived levels of autonomy on a scale, e.g., from 1 to 5 (Bradtke et al., 2016). Such a statement, for instance, for schedule autonomy, is the following: 'I have control over scheduling my work' (Breaugh, 1985). Then, the average of the perceived autonomy over each dimension and over different tasks defines the overall work autonomy (Green, 2008). Since it can only be roughly determined in advance, how much autonomy is actually given to employees and how much autonomy they precisely perceive, i.e., an employee's individual perceived level of work autonomy, is normally measured retrospectively.

Assessing the autonomy level of a technical system, on the contrary, is normally based on a criteria catalogue. The higher the capability level of a technical system, the more criteria of such a catalogue it fulfils, and thus it is classified as more autonomous. Therefore, the targeted autonomy level can be determined when a system is designed. More autonomy usually increases development efforts and costs. For industrial contexts, six levels of autonomy have been defined, reaching from no autonomy at all (0: meaning that a human has full control) through temporary autonomy (2: a human still defines goals) to full autonomy (5: fully autonomous operation of a system; a human does not have to observe this system) (Ahlborn et al., 2019). In their two-dimensional taxonomy, Simmler and Frischknecht (2021) connected automation and (system) autonomy by defining five levels for both concepts, considering human-machine collaboration. Their automation levels reach from decision support through human approval and information to full automation without human involvement. Regarding autonomy, they distinguished between deterministic, non-transparent, undetermined, adaptable, and open systems. Probably the most popular definition of autonomy levels concerns autonomous driving and is found in the international norm SAE J3016. This definition starts at level 0, which indicates that the technical system is not automated at all, and goes up to level 5, which indicates that the vehicle is fully automated and controls the driving regardless of the current situation and driving mode. While Bertschinger et al. (2008) tried to develop quantitative measures of system autonomy based on an information theoretic approach but concluded that the general autonomy of a system is not quantifiable, Beernaert et al. (2018) succeeded in defining a mathematical function of independence and task complexity that can be used to measure a system's autonomy. Still, Bertschinger et al. (2008) argued that only behavioural autonomy within a specific environment and for a defined task is measurable. Nevertheless, it might be easier to plan and design a particular level of autonomy for a technical system than to precisely determine human autonomy in advance.

2.3 Prerequisites for Work and System Autonomy

In general and when working with technical systems, work autonomy above all arises from employees being able to understand what they are working on, identify points at which they can influence work processes, understand how to use intervention points and work technologies, and establish new ways of accomplishing tasks (Deci et al., 2017; Friedman, 1977; Hacker & Sachse, 2014), which can be called situation awareness. This means that to make work-related decisions autonomously, employees need transparent work procedures, and they need to be able to predict the outcomes of their own actions or cooperation partners and be capable of controlling work processes (Hacker & Sachse, 2014). This presupposes that they have the knowledge, experiences, and adequate qualifications that make it possible for them to understand job demands and to be able to cope with them, including being able to use technical systems (Green, 2008). Employees can either acquire these prerequisites of work autonomy in specific training or learn them through experience. To avoid the loss of human knowledge and skills, employees rely on opportunities to regularly practice and develop these skills.

Due to the requirements of future manufacturing, autonomous systems can enhance flexibility and are also able to perform in unknown situations and environments (Endsley, 2017). Being self-reliant, autonomous systems need less interaction and parametrization. If autonomous systems moreover involve machine-learning capabilities, they can even learn from past experiences and adapt their future behaviour accordingly. Looking at the prerequisites for technical autonomy, fundamental abilities that are natural for humans have to be implemented in autonomous systems explicitly. Instead of programming the fixed and pre-determined behaviour of a system, the goal is rather to model capabilities in the form of skills. These skills can be hierarchically combined, strung together, and thus used in different contexts to build powerful and capable systems. A higher-level but crucial capability similar to that of humans is situation awareness. Perceiving the current environment, context, and situation with respect to a certain task or goal, and possibly a human counterpart, is a complex and challenging task for technical systems. Therefore, systems have to be equipped with a variety of sensors, and their computing units have to process the incoming data, use these data as the input for evaluation algorithms, and constantly update a world model, which is a fundamental requirement for autonomy. This world model could consist of several sub-models and serves as the knowledge repository of an autonomous system. In situations that require human-machine interaction, a distinct user model enhances the interaction quality, since it allows the personalisation of interaction procedures. In addition to interaction with humans, interaction with other systems is also often crucial for system autonomy. Accordingly, technical systems require standardised connectivity and interfaces for data exchange and negotiation.

Providing technical systems that are able to transparently communicate with their human colleagues can be challenging but is also important for human autonomy. A higher technical autonomy of these assistance systems requires better awareness of their working environment, which could indicate the need for sensors and monitoring systems. Comprehensive and permanent monitoring, however, might compromise human autonomy, as it puts stress on humans, causing them to feel restricted in deciding among different options for action (Stanton & Barnes Farrell, 1996).

2.4   Negotiation of Autonomy

The concept of the employment relationship implies that employees, managers, supervisors, and also co-workers represent the employer and its interests through interacting with other employees and controlling their performance (Coyle-Shapiro & Shore, 2007). Employment relationships are reciprocal commitments that rely on trust and loyalty (Cropanzano & Mitchell, 2005). This means that, like any other social exchange, these relationships follow a norm of reciprocity, where every party involved expects to get something (immediately or in the indefinite future) in return for their efforts and contribution. Employees, for instance, contribute their labour, skills, knowledge, and experiences (Pongratz & Voß, 2003; Böhle, 2018). Employers reward their employees with pay, prestige, and appreciation, but also grant them opportunities for self-fulfilment, self-monitoring, and different levels of autonomy. Within the limits and requirements of the job, like task interdependence, strictly predefined standardised work processes, or quality specifications, employees are able to negotiate for and claim rewards like autonomy. Thus, the negotiation process is central to how to distribute autonomy among employee groups. Implementing digital assistance shifts existing exchange and negotiation structures and practices, so that employees have to additionally assert themselves against digital assistance systems and negotiate on autonomy levels with these technical systems. Since successful work processes are built upon reliable and predictable collaborative actions for which the mutual expectations of the actors involved are met, infinite levels of autonomy would impair the collaboration. This makes work autonomy a resource that is not infinitely divisible.

Several scenarios of autonomy distribution in which autonomous digital assistance can additionally restrict or even enhance employees' autonomy are conceivable (for details, see the scenarios described in Sect. 4). Schilling et al. (2016) argued that shared autonomy between humans and technical solutions must be investigated; this includes the dynamic negotiation of autonomy degrees depending on the current context and situation. This shared autonomy concept also takes into account the fact that although a technical system might generally be classified as completely autonomous, in situations that require interaction with a human, the system might adapt its own autonomy degree.

Adapting their own autonomy is a complex task for technical systems that requires specific strategies and strongly relies on human-machine interaction. This can either be realised through a specific negotiation interaction between a human and the system, or the system might adapt automatically, perhaps without informing its user explicitly. Therefore, a technical system must be capable of reducing its own scope of action. If a system is used in cooperative scenarios, it needs an understanding of human capabilities and the functionality to sense and negotiate the actual task allocation. Furthermore, it is important that a system also communicates its own understanding of the shared autonomy to ensure agreement on a negotiation result. Research shows that transparency about a system's state is beneficial for interactions with human users and can even compensate for further technical shortcomings or maladjustments (O'Neill et al., 2020; Lyons et al., 2021).

In a recent interpretation of explaining the processes of AI systems, Rohlfing et al. (2020) emphasised that in order to make explanations successful, i.e., to reach an understanding, human users must not be seen as passive receivers of information but as active participants that shape the explanation process and thus participate in a co-construction activity. More specifically, Rohlfing et al. (2020) defined co-construction as a bidirectional interaction involving both partners that consists of constructing a task and its relevant (recipient-oriented) aspects, requiring the monitoring and scaffolding of both interaction partners. From a macro-perspective, the interaction thus constitutes a social practice; it is an established (but flexible) interaction pattern consisting of joint actions toward a goal that are performed in a sequence.

In consequence, Rohlfing et al. (2020) argued that such co-constructive activities require a dynamic partner model that takes the changing levels of understanding of an interaction partner into account in order to enable the negotiation of information. It is further argued that such an approach is required in order to empower users to actively shape aspects of an AI system and thus increase their autonomy.

However, explicit negotiations between a human and the system could also be disruptive if they happen too often. For small adaptations that the system is fairly sure will be accepted by the user, autonomously executed adaptations are an alternative option. Even within human-human interactions in work situations, autonomy is not always explicitly and consciously negotiated. When multiple autonomous systems have to negotiate their relative autonomy levels, systems can rely on simplified negotiation patterns. These patterns might be realised in the form of rule-based approaches that define a specific order of negotiation steps and criteria that determine shared levels of autonomy. For collaborations with increasingly autonomous technical systems (which at least partly take over employer functions), the question of whether and to what extent it can be ensured that employees continue to receive non-monetary rewards, such as autonomy for their contributions, arises. How employees are able to make legitimate claims for autonomy will likely have to be renegotiated, and what information assistance systems need to negotiate with must be determined. In this context, the system's reliability and to some extent transparency can contribute to enhancing employees' trust in their technical negotiation partner. These aspects are vital for establishing and protecting stable employment relationships. This also holds true for employees interacting with digital assistance systems.

3 Digital Assistance

In the section above, we contrasted human work autonomy and system autonomy, which emphasised their relevance in flexible working environments. We will continue by first introducing our definition and understanding of digital assistance. Then, in Sect. 4, we will demonstrate why future assistance systems will need autonomous features, and we will elaborate on how these features can be implemented in Sect. 5. The use of digital assistance is meant to make work processes more efficient, ensure occupational safety, and better integrate unskilled workers or workers with disabilities (Mark et al., 2021). In addition, the need for digital assistance systems was also induced by the development of automated production systems, which require human operators for monitoring, decision-making, and manual task execution, and demands for more individualised products with smaller lot sizes (Gil et al., 2019). Thus, assistance systems are used to monitor and maintain fully automated systems or to perform manual work processes in which collaboration with humans is indispensable. Therefore, humans play an important role when it comes to digital assistance systems. To better understand the relationship between human work autonomy and digitally controlled and/or assisted work processes, we first define digital assistance by specifying its central aspects and describe the state of the art. Furthermore, we distinguish between different kinds of assistance, as its design, implementation, and applications vary.

Although the term 'digital assistance' is used in many contexts, such as, for example, driver assistance (in cars and other vehicles), personal assistance (e.g., fitness trackers and apps), and speech assistance in our homes, we want to focus on work-related assistance in this chapter. For work contexts, Fischer et al. (2018) defined digital assistance systems as either software solutions or combinations of software and hardware with which people interact and that support them in their current work situation. Apt et al. (2018) characterised digital assistance systems in terms of the kind of support they provide (physical, sensory, cognitive), the goal of the support they provide (compensatory, preserving, extending), and their degree of support (low, medium, high, variable). Most of the currently implemented assistance systems are often specialised for one support goal and have a fixed degree of support.

There are many digital assistance systems for shop floor tasks: They range from providing recommendations through supporting decisions, describing necessary actions, supporting the worker by taking over parts of the work, and documenting the execution of tasks to delivering answers for specific questions and more. Gil et al. (2019) distinguished between three human roles that exist in automation scenarios: monitoring, decision-making, and execution. Today, different digital assistance systems support humans in all of these roles, as the following examples from the literature show: Monitoring assistance, as presented by Buchholz and Kopp (2021), can support users, for example, in observing a number of control values that must be kept within a certain range. By directing the users' attention and also taking over some parts of the task at hand, the system adapts to its users' current performance. Decision-making assistance supports human decisions with the help of intelligent algorithms (Töniges et al., 2017); for example, there are algorithms for the aggregation and visualisation of production data from an entire shop floor (Kuhlmann et al., 2018; Mühge, 2018) and algorithms that provide decision support in maintenance and repair scenarios (Ullrich, 2016). Execution assistance can be found on the shop floor in the form of digitally equipped assembly stations at which assistance systems instruct human workers performing assembly processes. These processes might be too difficult to automate or they might, for example, require human dexterity (Oestreich et al., 2019); they also may have other requirements for which human flexibility is more important than the benefits of automation.

4 Interactions Between Human Work Autonomy and Autonomous Digital Assistance

As the previous section described, when employees work with digital assistance, they repeatedly interact and share different (parts of) tasks. Consequently, employees and digital assistance systems coordinate work methods, schedules, and objectives, which could affect their autonomy. In recent years, the drawbacks of what we would call static digital assistance, or simple digital copies, as Fischer et al. (2018) called it, have become obvious due to their application in industry. Evidence shows that digital assistance as it is implemented in work organisations often somewhat restricts employees' work autonomy (Butollo et al., 2018; Kellogg et al., 2020; Wood, 2021). A study on a particular digital assistance system that provides work instructions to employees shows that employees perceive that they have less work autonomy compared to employees who never work with this system when the digital assistance system pervades employees' activities in such a way that they work with it daily (Gensler & Abendroth, 2021). Then again, digital assistance is designed to 'assist' human employees and thus can encourage them to autonomously apply their skills and knowledge in more complex tasks. However, it is not technology alone that determines workers' perceived autonomy.

   When considering the following scenarios and examples of possible relationships between digital assistance and employees' perceived work autonomy, we must keep in mind that assistance systems differ in their properties and are embedded in varying organisational contexts, which further shape which of these properties have been adapted or are actually used (Orlikowski, 2000). Moreover, employees' use of digital assistance, skills, job resources, and competencies, as well as their '[...] assumptions, and expectations about the technology and its use, influenced typically by training, communication, and previous experiences' (Orlikowski, 2000, p. 410), affect how employees perceive digital assistance and how this is related to their work autonomy. Moreover, it is also possible that the assistance technology alone does not change how much work autonomy employees perceive. Certainly, the most interesting cases are those in which digital assistance does influence work autonomy and from which we can learn how to avoid restrictions or how to provide support to enhance employees' work autonomy.

The following examples from the literature show three different conceivable scenarios of the interaction between human and system autonomy in human collaborations with digital assistance. These scenarios are visualised along the x-axis on a 3×3 matrix in Fig. 1. In the first scenario, increasing autonomy on the system side leads to decreased autonomy for the worker. In the second scenario, increasing the system autonomy of a digital assistance system has no impact on workers' autonomy, and the third scenario describes an enhancing effect, where worker autonomy increases as system autonomy increases. The y-axis of the matrix indicates the three dimensions of human work autonomy (method, schedule, criteria). Since increasing the autonomy of digital assistance might only impact one specific dimension

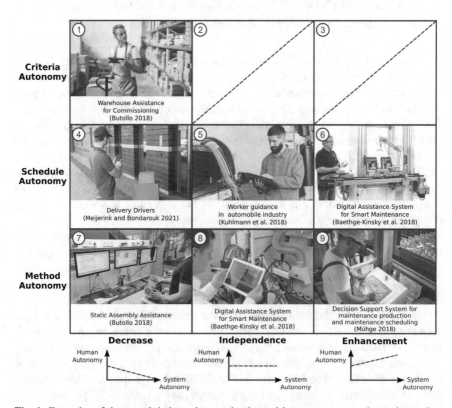

Fig. 1 Examples of decreased, independent, and enhanced human autonomy due to increasing system autonomy, classified according to the three work autonomy dimensions

of human work autonomy and could have the opposite or no effect at all for the other dimensions, distinguishing, in the three scenarios, between the three autonomy dimensions makes it easier to illustrate varying scenarios with different examples. The circled numbers in each cell make it easier to reference the combinations of scenarios and work autonomy dimensions. Each picture within the figure visualises a typical work and assistance situation from the referenced literature.

1 Decrease in Criteria Autonomy: A higher level of autonomy of digital assistance requires situation awareness; sensors are applied for this purpose. The assistance system gathers data and information about its work environment and also about employees and their performance. Employees could perceive this as surveillance and feel that their criteria autonomy is restricted. This is because digital assistance systems make it possible to immediately determine (and sanction if necessary) whether and to what extent employees have met work objectives by gathering data about their performance. Another example of low criteria autonomy at a high autonomy level of digital assistance is the use of data glasses in warehouse-commissioning scenarios. These data glasses automatically obtain the current work-related information of job management systems, which they use to record productivity data, movements, and the interactions of employees (Butollo et al., 2018). Based on that information, they are able to generate work objectives and calculate performance scores. Additionally, the Mentor App used by Amazon for its delivery drivers has also been found to gather driving and performance data to autonomously evaluate work objectives, thus putting employees' criteria autonomy at risk (Peterson, 2019). The implementation of the app and its monitoring feature had been justified by improvements in employees' work and road safety. In addition, inadequate communication, employees' lack of self-determination concerning their personal data, and the non-transparent intended usage for instant performance evaluation by managers have resulted in decreased acceptance of the technical system and criteria autonomy (Backhaus, 2019).

2 + 3 Independence and Enhancement of Criteria Autonomy: We could find no specific example in the literature that explicitly shows an interaction between system autonomy and human autonomy that had no impact on or increased the criteria autonomy. This might indicate a research gap in the area and is an interesting topic for future investigations.

4 Decrease in Schedule Autonomy: Digital assistance systems can reduce employees' schedule autonomy because these systems can specify new sequences and change accustomed working paces (Elliott & Long, 2016). An example of warehouse commissioning shows that to enhance efficiency in the labour process, using algorithms, technical systems can autonomously capture workers' operations and performance (Delfanti, 2021). This is necessary for allowing the digital assistance system to compute and reorganise warehouse storage. The changed warehouse organisation aims to efficiently use space and integrate cooperative work, resulting in 'chaotic storage', which is opaque from a human perspective. The more autonomous technical systems in this context organise work according to their algorithm-based storage system; thus, more employees depend on digital assistance, which guides them and schedules work. Similarly, Meijerink and Bondarouk (2021) reported the negative impacts of the app-controlled management of delivery and Uber drivers on their schedule autonomy since either tasks are automatically allocated to them or choices regarding upcoming tasks are filtered based on unknown criteria.

5 Independence of Schedule Autonomy: One example of digital worker guidance in the automobile industry (Kuhlmann et al., 2018) indicates that the implementation and the properties of digital assistance make a difference in how its use is perceived by employees in terms of their autonomy. Though digital worker guidance specifies fixed orders, employees are still able to choose their own work rhythm and perform several individual operations at a time instead of confirming each step, for example. On average, this should neither decrease nor increase the schedule autonomy significantly.

6 Enhancement of Schedule Autonomy: Digital assistance can be used to make work processes more transparent (Schildt, 2017) so that employees perceive that they can better control work situations. By communicating relevant, specific, and current information about work processes, digital assistance enables employees to better understand the decisions being made by colleagues, supervisors, or even technical systems. As a result, employees have a better chance to keep track of work processes and plan their personal schedules autonomously and in advance. An example of digital assistance for maintainers shows that technical autonomy can thus enhance schedule autonomy. In this example, the digital assistance system reports faults to the maintainers so that they can access information about the tasks to be performed (e.g., videos or photos of the machine parts to be repaired) anywhere in the plant (Baethge-Kinsky et al., 2018). This example also indicates that if employees are involved in complete work procedures and receive an overview of their job demands from the digital assistance system, they can intervene more successfully and apply their skills to schedule work autonomously. Hence, when employees interact with digital assistance, it is important that they are kept in the loop (Parker & Grote, 2020). Then, employees can better oversee single work steps and understand the degree to which they can influence work.

7 Decrease in Method Autonomy: Using digital systems often means decomposing work into many single working steps and thus introducing standardised procedures (Ittermann & Niehaus, 2018). Standardised and small-scale work steps imply more routine operations that require less specific skills or knowledge and thus less individual scope for action. Moreover, decomposing work can result in human workers being out of the loop, as they miss interim steps and cannot always fully reconstruct how decisions are being made (Delfanti, 2021; Parker & Grote, 2020). The example described by Butollo et al. (2018) also indicates this relationship. To enable the assistance systems to have more autonomy, activities are standardised and work tasks are decomposed into single and simple tasks. This results in routine tasks for employees, and there is little variation in how tasks are to be performed. Employees then hardly have any opportunity to develop their own methods for task execution, which restricts method autonomy.

8 Independence of Method Autonomy: In some cases, the occupational status given by the skills and expertise ascribed to employees moderates the effect of the use of digital assistance. Baethge-Kinsky et al. (2018) explained that a studied maintenance assistance system does not seem to have any effect on the maintainers' method autonomy. However, in their study, the assistance system has little autonomy and is mainly used for paperless documentation and mobile spare parts management. For both functions, the system depends on workers' input. Moreover, these employee groups already have high autonomy levels and consider themselves essential; thus, they have a high degree of self-assurance. This self-assurance seems not to be rattled by the implementation of a digital assistance system that is associated with changes in the organisation of work.

9 Enhancement of Method Autonomy: In some cases, digital assistance enables employees to make responsible decisions more autonomously by referring to detailed work-related information. In addition, an assistance system that includes monitoring functions can reduce the risk of mistakes, as they are detected in real time, and thus it can enhance method autonomy. This function contributes to better working conditions in which employees experience more personal responsibility, even in more complex or risky work situations (Kuhlmann et al., 2018). In addition, employees do not have to interrupt co-workers or supervisors to receive help, which enables them to make more self-determined decisions on how to do their work. Workers are moreover empowered to follow a self-directed learning path if the assistance systems include analytic capabilities for tracking skill development and certification mechanisms to show that workers have completed courses (Apt et al., 2018). According to Mühge (2018), the introduction of decision support systems in manufacturing changes the work of mid-level-management employees since they receive more non-repetitive and non-rule-based tasks, which increases their method autonomy. Through the implementation of digital assistance, employees should be able to gain more knowledge on varying tasks and acquire extensive skills, which will enable them to be more flexible and autonomous in performing different and also more complex tasks.

As illustrated, various interaction scenarios between human work autonomy and technical autonomy are conceivable. Therefore, the following question arises: To what extent can a system design deliberately impact the work autonomy of its users? This topic has increasingly been investigated in recent years. Adaptive assistance should allow personalisation and contextualisation to ensure that despite the rising autonomy levels of technical systems and particularly assistance systems, human beings are kept in the loop and the positive potentials of technical autonomy are used to support humans in their daily work. Thus, it can be assumed that adaptive assistance can directly affect human work autonomy.

5   Adaptive Assistance for Improved Human Work Autonomy

While current digital assistance systems mostly act as digital copies of paper-based instructions with some advancements (Fischer et al., 2018), future adaptive digital assistance systems will adapt to their users, the current situation, and the work environment. Hence, Morana et al. (2020) claimed that an intelligent system needs human- , task-, and/or context-dependent assistance and capabilities to augment its interactions with users.

If the negative impacts of digital assistance are considered as an opportunity to learn and improve, and positive impacts are reinforced, adaptive digital assistance systems can be developed that enhance work autonomy for employees. It appears that digital assistance restricts work autonomy when it is inflexible or, for example, does not take its human user into account by transparently communicating its actions and reasoning. Adaptive digital assistance systems, by contrast, are able to better consider their human counterpart in their decisions concerning (a) how to support the human, (b) how to communicate their own behaviour and adaptations to the human, and (c) how to negotiate responsibilities in task execution to establish so-called shared autonomy (Schilling et al., 2016). This suggests that the adaptivity of digital assistance can contribute to enhanced human work autonomy. Although human autonomy is rarely discussed explicitly by researchers in the technical domain, adaptive features of assistance systems can contribute to enhanced work autonomy and more flexibility by sharing autonomy between systems and employees.

O'Neill et al. (2020) coined the phrase 'human-autonomy teams' to describe situations in which two or more agents work interdependently toward a common goal (and at least one agent is a computer entity). By introducing the term 'socio-technical assistance ensemble', Selke and Biniok (2015) emphasised that there are always two sides of assistance: One side involves offering assistance and the other involves receiving assistance, with a shared responsibility and the goal of accom-plishing predefined tasks in a cooperative manner. If assistance becomes adaptive, it acts more like a companion than an instructor (Haase et al., 2020). Wendemuth et al. (2018) stated that companion technology complements the technical functionalities of an intelligent system to enhance its interactions with users, and combining these two technologies can create a cooperative and trustworthy relationship between the user and the system. This can be achieved by introducing more adaptivity into assis-tance systems. When these systems become more flexible by adapting to the current situation and context and by providing personalisation and individualisation, digital assistance becomes partially autonomous. With an increased level of autonomy, a technical system is perceived to be more like a co-worker instead of simply being considered a tool (O'Neill et al., 2020).

Distinguishing between adaptive and adaptable systems is especially important when discussing autonomy (Burggräf et al., 2021). Following the definition of Loch et al. (2018), in adaptable systems, it is up to the users to change the system, while in adaptive systems, the system itself initiates changes. At first glance, it may seem as if adaptable systems offer more human autonomy since users define which adaptations happen and when. However, adaptable systems introduce the risk that users may not even be aware of all customisation options and therefore are unable to configure and parameterize the systems according to their needs. Adaptive systems, in contrast, determine the kind of adaptation that is applied and the moment when it is executed. This might decrease the perceived human autonomy, and this underlines the need for the high transparency, good explainability, and good negotiability of the system.

Adaptive features of digital assistance systems can pursue different goals like providing enhanced ergonomic support, more efficient execution, or guidance conducive to learning new tasks and skills (Oestreich et al., 2022). Adaptivity in assistance systems has different forms and features: The system might use different modalities for user interaction depending on the situation. It might switch from a graphical presentation of the instructions to delivering them through speech synthesis (Josifovska et al., 2019). Another option is that a system might adapt the presented type of information depending on its user, e.g., by showing more videos, images, or texts. A rather advanced adaptation is the modification of the underlying content structure. By doing this, a personalised learning path through the teaching material can be put into practice (Oestreich et al., 2021).

Adaptive assistance systems, however, create further challenges compared to the usual digital assistance systems. Assistance systems usually perform tasks or solve problems according to previously programmed specifications (Alpaydin, 2016). This may already exceed human cognitive capacities, as these systems reason in a way that significantly differs from human modes of thought (Dourish, 2016). However, this especially holds true when algorithms that automatically learn and adapt by considering new information are applied (Burrell, 2016). As a result, tasks supported by or decisions being made by adaptive digital assistance systems can become opaque (Burrell, 2016), and humans are at risk of losing track of the current work situation and becoming even more dependent on digital assistance (Delfanti, 2021).

Nevertheless, advanced digital assistance systems can provide personalised and essential information on work processes when they are adapted to current work situations (Kuhlmann et al., 2018; Mühge, 2018). There are manifold adaptive features of assistance systems, some of which seem trivial: Adapting the workplace height to ergonomically suit the individual, adapting font sizes and colour themes to compensate for disabilities, and adapting the instruction depth/detail to the experience level of the user are all tasks that can be carried out by assistance systems (Yigitbas et al., 2020). The degree to which these adaptive features affect work autonomy depends on their concrete application and the work task, situation, and environment.

In addition to the ability to adapt support in a personalised manner, there are several prerequisites that are necessary to the autonomous execution of adaptations in a digital assistance system. First of all, an adaptive system needs some kind of situational awareness to detect when adaptation is necessary and whether it is the right time to execute an adaptation (Loch et al., 2018). Making sense of incoming data requires sensors and intelligent processing. These data are then combined and filtered to create a world model that an autonomous system can use to determine and plan its actions. Part of this world model is also a distinct partner model. This dynamic user model has been recently discussed in the literature as the digital twin of the human (Josifovska et al., 2019; Engels, 2020). Second, the system needs the ability to negotiate responsibilities with its human counterpart. Finally, an adaptive system must be able to transparently communicate automatic adaptations to ensure that they are visible to and controllable by its user. In interaction situations in which people need to be able to take over the execution of a task, it is important that they are kept in the loop, which means that a system must continuously inform the user about the current state of the execution of a task. Furthermore, in cooperative scenarios in which humans and systems perform a task together, the task allocation must be flexible enough to ensure that a system can relieve the human if necessary, but human workers still must keep their capabilities and skills, and must participate in regular training.

In the previous section, we have provided an overview that shows which aspects should be considered when adaptivity and human work autonomy are discussed. In the following, we present a guideline to improve human-system interaction through the use of adaptive features, with a particular focus on the distribution of autonomy between the employee and the digital assistance system.

6 Guideline for Enhanced Work Autonomy Through Adaptivity

Because the aim of this contribution is to raise awareness of human work autonomy as an important design factor for digital assistance in general, we give some concrete design suggestions for digital assistance to encourage stakeholders to consider human autonomy more in the design processes of digital assistance systems (see Table 2). While Beernaert et al. (2018) chose a technology-centred approach and came up with two design principles to increase system autonomy, we want to motivate a human-centred guideline, which should enable or even enhance human autonomy in interactions with assistance systems that are becoming ever more autonomous. Our suggestions are the result of an analysis of the negative and positive impacts of current assistance system approaches and our review of current and future approaches in the literature. Furthermore, they are the result of our own experiences and insights from several research projects (e.g., Oestreich et al. (2019, 2021)) and interdisciplinary discussions within the research programme 'Design of Flexible Work Environments— Human-Centric Use of Cyber-Physical Systems in Industry 4.0', in which relevant previous results have also been published (Gensler & Abendroth, 2021; Oestreich et al., 2020, 2022). The following list of recommendations serves as a first step in initiating discussions about human work autonomy in interaction with digital assistance. Therefore, this list is not meant to be complete, applicable to all systems, or implemented completely. We categorised the design recommendations in Table 1 according to the method, schedule, and criteria autonomy dimensions, which we consider to be particularly important within the collaboration between employees and digital assistance. As Table 2 reveals, the design and implementation of digital assistance systems do not affect each work autonomy dimension equally. The most affected dimension seems to be method autonomy (1.1–1.7), which is followed by schedule autonomy (2.1–2.3), and the least affected dimension seems to be criteria autonomy (3.1).

However, one challenge is that there is no ideal level of work autonomy for all users, all tasks, and all situations. Too much autonomy, for example, can leave

Table 2 Guideline for the work-autonomy-enhancing design of assistance systems

#	Design recommendations
1.	**Method autonomy**
1.1	**Decrease assistance over time for repeated tasks**
	Recommendation: Relying on digital assistance for a longer period of time leads to de-skilling and thus should be avoided
	Explanation: Focusing on the method autonomy of workers, our own research (Oestreich et al., 2019) and the research of others (e.g., Gräßler et al. (2020)) has shown that users learn over time and thus, a decrease in assistance is necessary to ensure the acceptance of a technical system and to avoid de-skilling
1.2	**Keep interfaces, interactions, and procedures as adaptable tasks**
	Recommendation: Regardless of whether the individualisation of a digital assistance system is automated (adaptive) or manual (adaptable), a system should be able to adapt to its current user and their preferences to enhance human autonomy
	Explanation: Adaptability and adaptivity are considered two separate ways to personalise a system and its behaviour (Burggräf et al., 2021; Loch et al., 2018; Bunt et al., 2004). Adaptability ensures that the human worker stays in control and is able to undo adaptive changes in the system
1.3	**Explain autonomous behaviours**
	Recommendation: To ensure human autonomy, it is important for the human to understand and anticipate system behaviour to a large extent. If a system adapts its own behaviour, because it has learned or detected the necessity to do so (e.g., rule-based), this must be communicated to its user. This ensures trust when humans closely interact with a system
	Explanation: Regarding a system's ability to communicate about itself and its behaviours, Weitz et al. (2021) claimed 'that a lack of transparency, [...], might have a negative impact on the trustworthiness of a system', and the review study of O'Neill et al. (2020) described similar findings
1.4	**Provide contextualised, precise, and the most necessary relevant information**
	Recommendation: Support users in unknown situations and empower them to keep acting autonomously; in addition, help them avoid the need to ask colleagues for help and reduce the risk of overwhelming them
	Explanation: Supporting users in unknown or at least unfamiliar work situations is one of the main reasons for the implementation of assistance. Since interacting with a technical system sometimes represents a lower barrier than asking a colleague or friend for support, the assistance system itself might be autonomy-enhancing for users
1.5	**Offer decision options and advice**
	Recommendation: Keep the decision-making power on the human side but provide support to allow the human to make an informed decision
	Explanation: Visibility, controllability, and explainability are important parameters that define how an adaptation is presented to a user. As Bunt et al. (2004) reported, some users simply prefer to stay in control of their accustomed user interfaces

(continued)

Table 2 (continued)

#	Design recommendations
1.6	**Observe task execution using advanced sensors and intelligent processing** *Recommendation:* Ensure that no errors occur and track the execution of a task without necessarily logging the personal information of an employee. A user should feel supported by a system, not controlled or even punished for errors *Explanation:* Intelligent assistance systems are capable of monitoring the execution of tasks and automating quality checks. The challenge is to make users feel supported by the system rather than controlled. Since a hint on how to perform a task from a technical system might be less embarrassing than a hint from a co-worker or supervisor, we think that a technical solution designed in the right way could even support workers' autonomy. As Heinz and Röcker (2018) mentioned, data security and workplace privacy must always be considered
1.7	**Explicitly negotiate autonomy levels between humans and systems** *Recommendation:* For enhanced human autonomy, it is important to know who is responsible in a specific situation; in situations in which the human is in charge, the system should make the human feel safe and avoid errors *Explanation:* It is important for an advanced autonomous system to act more like an intelligent agent and less like an inflexible tool (O'Neill et al., 2020). This helps to increase humans' trust and even promotes autonomy if humans are able to intervene in decisions made by the system. Co-construction, as described by Rohlfing et al. (2020), allows humans to use more technical support when it is needed but also decreases and minimises assistance when workers feel secure and competent
2.	**Schedule autonomy**
2.1	**Transparently communicate and indicate automated procedures and the system status** *Recommendation:* Keeping the human in the loop is necessary to ensure situational awareness and to allow the human to interfere and take over tasks if needed *Explanation:* While the problem of nontransparent communication and an unknown system status is a special problem in, e.g., driver or monitoring assistance, it is also relevant for workers using digital assistance in other industrial environments (Heinz & Röcker, 2018). O'Neill et al. (2020) reported mixed effects regarding the transparency and communication of autonomous systems that cooperate with humans. It is important to meet an adequate level of transparency and to avoid making excessive demands of workers in situations that involve a high cognitive load. However, information asymmetries must be avoided since they have negative effects on worker autonomy, as mentioned by Meijerink and Bondarouk (2021)
2.2	**Provide freedom of choice regarding the order in which tasks and different orders are processed** *Recommendation:* As long as external conditions (priorities, safety protocols, task interdependence, etc.) allow it, users should be able to choose the order in which they complete tasks *Explanation:* Butollo et al. (2018) reported that the usage of digital assistance systems often enforces standardised work procedures and therefore limits human schedule autonomy. Similarly, Lammi (2021, pp. 122–123) pointed out the inability of employees to plan work when automated work delivery systems are employed. This may conflict with personal habits and periods of concentration and productivity. Therefore, restrictions related to the order in which tasks must be completed should be minimised

(continued)

Table 2 (continued)

#	Design recommendations
2.3	**Provide information about work goals and subgoals**
	Recommendation: Allow the user to plan single work steps and the total duration in advance
	Explanation: One pedagogical approach that can be used to encourage learners is to set goals and subgoals and to communicate those goals to learners as they interact with a learning support system (Woolf, 2009). With regard to schedule autonomy, users should be informed in advance about what tasks they will have to perform and in which order they ideally should be performed so that they can create self-determined schedules
3.	**Criteria autonomy**
3.1	**Provide possibilities for intervention and voice**
	Recommendation: When a task is completed and human performance is automatically rated, this allows the self-assessment of the work quality and work performance
	Explanation: When assistance systems judge operators' performance or automatically delegate tasks based on worker profiles, it is very important that humans are able to intervene and provide feedback on automatically calculated skill levels or similar measures. Graessler and Poehler (2018) considered this topic by allowing the declination of tasks delegated by a decentralised control system to a specific human worker

employees feeling overwhelmed and increase feelings of stress, which then lower their performance (Langfred & Moye, 2004; Kubicek et al., 2014). According to Sankaran et al. (2021), decreased human autonomy becomes especially critical if a motivational state called reactance becomes active. In this case, users perceive a threat to their personal freedom and invoke the need to reassert free behaviour. Therefore, shared autonomy between human beings and systems has to be distributed and negotiated continuously during the usage of assistance systems. However, this is a separate topic that should be discussed by the research community in the future. In addition, other recommendations go beyond the realm that digital assistance systems themselves can impact. For example, by installing digital assistance systems at assembly workstations throughout the shop floor, companies gain the ability to promote the job rotation of their employees. This would allow workers to change their tasks regularly to avoid executing the same routine work everyday, and they could even qualify for new tasks by using the provided assistance systems. If this system was then controlled by a higher-level digital control instance, as presented by Graessler and Poehler (2017), human work autonomy would be at risk, at least if no possibilities for human intervention were provided.

The concept of the digital twin of a human could play an important role in the implementation of some of the design recommendations (e.g., 1.1, 1.2, 1.4, 1.7). Therefore, in the following section, we want to critically reflect on its role in adaptive assistance and human work autonomy. We investigate which potentials are inherent in the use and implementation of the concept and where it has real added value compared to known concepts for user models.

7 Role of the Human Digital Twin in Work Autonomy

Originating in the context of manufacturing and product development, the term 'digital twin' has gained popularity in recent years. The digital twin is a virtual representation of its physical counterpart that makes it possible to document the whole life cycle of a product, run simulations, and actively interact with its physical equivalent. Rosen et al. (2015) emphasised the importance of a digital twin as a prerequisite for developing more autonomous production systems for future manufacturing systems. They claimed that the digital twin goes beyond being a big collection of artefacts and note that the structure and semantics are crucial features.

Due to the importance of human-machine interaction in socio-technical systems, the term has recently been adapted for humans, leading to the term 'the digital twin of a human'. An up-to-date model of a user, including details about individual long- and short-term factors (knowledge, skills, experience, personal needs) and the current user state (physiological reactions and behavioural reactions) (Schwarz et al., 2014), can be used to allow digital assistance systems to adapt to their users. In the domain of digital assistance, to our knowledge, the term was first introduced by Graessler and Poehler (2017). The concept of the digital twin was then further developed by Graessler and Poehler (2018), who presented the idea of using the digital twins of humans to schedule personnel for workstations, based on different factors like capabilities and availability. Focusing on contextualisation and personal adaptation, Josifovska et al. (2019) presented a user interface adaptation framework for digital assistance based on the digital twin of a human. Since adaptation to learning and knowledge is a crucial feature of successful and satisfying assistance (Fletcher et al., 2020), the digital twin concept can be used in this context to store all knowledge about task executions and interactions with a digital assistance system. Learning with a digital twin as a companion has also been discussed by Berisha-Gawlowski et al. (2020). In addition to simply storing historical interaction data, the digital twin can also be used to learn from the gathered data to personalise an assistance system for its user. A predictive model could be used for this purpose, and according to Eramo et al. (2021), it should be a part of the digital twin architecture. This way, interactivity between the system, the human being, and the human's twin can be implemented. However, future research must investigate various questions related to the use of such a detailed digital representation of a human. For machine use, data are often aggregated and converted into formats that are no longer understandable by humans. Information that is difficult for humans to interpret should thus be avoided to increase acceptance and maintain transparency. Furthermore, adaptive systems gain the potential to simulate and negotiate adaptations implicitly with the digital twin of a human. While this might enhance the adaptation quality and thus personalisation on the one hand, on the other hand, engineers should be aware of the risk of the human being out of the loop and losing the power to explicitly control the system. In addition, the standardisation of the concept of the digital twin of a human would allow reuse in multiple systems and versatile contexts (Zibuschka et al., 2020). Another idea would be to incorporate the digital twin of a human into the digital twin of a complete

socio-technical system. This might allow one to evaluate how autonomy is distributed between a technical system and its user. Furthermore, this could make it possible to more efficiently adapt autonomy levels to current work situations, job demands, and not least to user states (e.g., situation awareness, workload, engagement, etc.).

However, risks and concerns regarding surveillance, data protection, and privacy must be taken seriously and are often regulated by works councils (Baumann & Maschke, 2016) or governments (Varadinek et al., 2018). To create digital twin models of their human users, assistance systems monitor, gather, store, and analyse employees' behaviour, which can affect humans' perceptions of their work autonomy (Tomczak et al., 2018). Perceiving workplace surveillance increases the risk of employees feeling stress and job strain, and it can reduce their job satisfaction (Backhaus, 2019). What kind of data is stored, where the data are stored, who has access to the data, and what the data are used for should be transparent (Backhaus, 2019; Zibuschka et al., 2020). Perhaps, in advanced versions, humans could even interact with their own twin to update or erase data, or at least to inspect which data are stored.

There are several additional factors that affect the way people perceive the collection and storage of digital data about themselves; this will be discussed in more detail hereinafter. First, humans feel more restricted in their work autonomy when they are single objects of observation than when the monitoring of the team's behaviour is performed (Baethge-Kinsky et al., 2018). Second, someone who is monitored permanently feels more restricted in their autonomy than someone who is monitored sporadically (Backhaus, 2019), even if permanent monitoring improves personalisation and adaptivity in general. Random samples of workplace performance monitoring, however, can also create stress and restrict autonomy when users are not informed about at which time points and how regularly they are being observed. Third, enabling users to access, change, and determine the use of their data increases the acceptance of workplace monitoring and thus employees' feeling that they are still in control of their own data (Backhaus, 2019; Tomczak et al., 2018). Fourth, if the use of a digital twin means that assistance systems are used to both assign and control tasks, this restricts employees' perception of autonomy more than if only one of these functions is used (Backhaus, 2019). Further research in the context of this project suggests that employees do not feel restricted in their work autonomy per se when they receive automatic instructions from a digital assistance system and also when data and information about work processes are automatically stored. The latter can be described as algorithmic monitoring and contributes to the creation of the digital twins of humans. Nevertheless, in any case, employees' occupational expertise should not be devalued, so that they still feel able to perform their work autonomously, when such monitoring is used to create the digital twin of a human. Fifth, the impact of single digital assistance systems that could both assign tasks and monitor employees' behaviour is contingent on the communication about these systems and their implementation (Tomczak et al., 2018). Employees who are able to participate in the implementation processes of digital assistance systems feel less restricted in their work autonomy than employees for whom all these processes are opaque and not controllable (Backhaus, 2019; Tomczak et al., 2018).

To summarise, these findings suggest that stakeholders should strive for a good balance between the intended use of the digital twins of humans and unnecessary data collection, since such comprehensive user models can either enable the support of and adaptation to humans (Evers et al., 2018; Oestreich et al., 2021) or create job stress and feelings of mistrust by restricting work autonomy. In the design and implementation of adaptive digital assistance, the model of the digital twin of a human could help to improve adaptivity to humans' needs and capabilities, because it enables the simulation of various work situations in short amounts of time without affecting work procedures or distracting humans from their tasks. However, if the digital twin of a human is used for this purpose, engineers should consider and analyse carefully which data and observations are truly necessary for this process.

8   Conclusion

This chapter aimed to make four central contributions to understanding the work autonomy of humans and technical systems and how adaptivity can help to reconcile both for the benefit of humans. First, we suggested a theoretical concept involving human and system autonomy in work processes. In doing so, we linked both concepts from an interdisciplinary perspective; they had mostly been discussed in a rather isolated way within their closely linked disciplines. Comparing and elaborating central aspects of human work autonomy and system autonomy enabled us to identify that both employees and digital assistance systems require, for example, knowledge and information regarding the other's intentions, skills, and actions, as well as their work environment, to be able to act autonomously. We also concluded that depending on individual situations and work conditions, humans and technical systems have to negotiate or 'co-construct' (Rohlfing et al., 2020) their respective relative levels of autonomy.

Second, to make the theory more tangible, we discussed the central consequences of current digital assistance systems for the shop floor for work autonomy and extracted the positive and negative impacts on human work autonomy. We emphasised that increased system autonomy affects the method, schedule, and criteria work autonomy dimensions differently. Therefore, we have argued that these dimensions of human work autonomy should be considered separately in the design of digital assistance systems. Based on the theoretical concept suggested before, our theoretical contribution underlined that the interaction between human work autonomy and autonomous digital assistance systems can result in decreased human work autonomy, the independence of human and system autonomy, or enhanced human autonomy. Reflecting particularly on those cases in which using digital assistance clearly affects employees' perceived work autonomy enables us to learn how to reduce restrictions, as well as how to improve employees' work autonomy. This is why we have discussed the use of adaptivity features to better address and enhance human work autonomy in the design and implementation of digital assistance systems. The adaptivity of digital assistance is vital, as employees, their individual needs and skills, and also work demands vary between tasks and workdays.

Third, we developed a guideline containing eleven recommendations to be considered in the design and implementation of adaptive assistance systems in the future. These recommendations are based on approaches for adaptive assistance and insights from relevant research findings. Our proposed guideline can be applied in concrete assistance scenarios. Two central aspects of the guideline are the following: First, one must ensure that users who work with digital assistance systems are kept 'in the loop' concerning work procedures at all times. Second, one should use adaptive assistance for the personalised support of employees, which enables work autonomy and thus the application and development of individual skills and knowledge. Admittedly, the suggested guideline still need to be evaluated. To do this, more research that tests the recommendations in practice is needed.

Fourth, we have discussed to what extent the concept of the digital twin of a human can contribute to improved adaptivity, focusing on the human need for autonomy and support in challenging work situations. We have identified limitations; the prerequisites for such a digital twin may infringe on employees' rights concerning their own data and privacy. Limitations also occur when employees are not able to access when and which data are automatically stored for the purpose of creating their digital twins, especially when they lack opportunities to change or erase these data. Then, employees perceive monitoring, data collection, and analyses as surveillance, and excessive data gathering will restrict their work autonomy and prevent them from freely developing and applying skills and knowledge.

Nevertheless, we believe that the digital twin of a socio-technical system has the potential to evaluate human and technical system autonomy, as well as their interaction. In this case, the digital twin can represent a good simplified depiction of employees' individual skills, needs, experience, current states, and work environment. Furthermore, the digital twin then implicitly simulates work processes without interrupting its human counterpart, although crucial interim steps and adaptions should be explained afterwards. This can improve personalisation, provide more interaction possibilities for the digital assistance system, and also improve transparency and explainability.

It would allow people to use the digital twins of humans as learning and training companions. Overall, the concept of the digital twins of humans provides a basis for further discussions and raises awareness of the importance of considering human capabilities and needs in designing and implementing digital assistance systems for work purposes.

Changing the working conditions promoted by digital assistance systems poses new challenges for the humane design of work. Still, from a social science point of view, we must acknowledge that more autonomous technical systems not only risk restricting human work autonomy but also have the potential to support employees in their work in a more targeted way without devaluing or even replacing them. However, it is crucial to determine how and in which work contexts digital assistance systems should be implemented: Permanent monitoring by digital assistance systems and the additional devaluation of one's own occupational qualifications, for example, could counteract the intended assistance and diminish work autonomy. Therefore, we need to further investigate what social and technical factors in the work environment allow employees to feel self-determined when working with adaptive digital assistance.

From a technical perspective, data models must be investigated to determine which data are actually needed for successful adaptations. How can systems transparently communicate their state and adaptations, and what are efficient and natural negotiation strategies for establishing a satisfying shared autonomy in work contexts?

Interdisciplinary exchange can help to create a better understanding and raise awareness of the limits and possibilities of new digital technologies like adaptive digital assistance systems. In addition, the joint discourse of the technical and social science perspectives can provide insights into the unintended negative consequences of these systems and solutions to address these consequences.

Author Contributions Conceptualisation: E.G, H.O., A.-K.A., S.W. and B.W.; Writing - Original Draft: E.G. and H.O.; Writing - Editing: E.G., H.O.; Visualisation: H.O.; Review and Supervision: A.-K.A., S.W. and B.W.; Funding Acquisition; A.-K.A. and B.W. All authors have read and agreed to the published version of the chapter.

Acknowledgements Images 1, 4, 5, and 9 in Fig. 1 were designed by Freepik.com, and images 6, 7, and 8 were provided by CoR-Lab, Bielefeld University. Elisa Gensler, Hendrik Oestreich, Anja-Kristin Abendroth, and Britta Wrede are part of the research programme 'Design of Flexible Work Environments—Human-Centric Use of Cyber-Physical Systems in Industry 4.0', which is supported by the North Rhine-Westphalian funding scheme 'Forschungskolleg'. All authors are affiliated with the Research Institute for Cognition and Robotics (CoR-Lab), Bielefeld University.

References

Ahlborn, K., Bachmann, G., Biegel, F., Bienert, J., Falk, S., Dr-Ing Fay, A., et al. (2019). Technologieszenario Künstliche Intelligenz in der Industrie 4.0.

Alaieri, F., & Vellino, A. (2016). Ethical decision making in robots: Autonomy, trust and responsibility. Lecture notes in computer science. In A. Agah, J. J. Cabibihan, A. M. Howard, M. A. Salichs, & H. He (Eds.), *Social robotics* (Vol. 9979, pp. 159–168). Cham: Springer International Publishing. https://doi.org/10.1007/978-3-319-47437-316

Alpaydin, E. (2016). *Machine learning: The new AI*. MIT Press essential knowledge seriesCambridge (Massachusetts): MIT Press.

Ansari, F., Hold, P., & Sihn, W. (2018). Human-centered cyber physical production system: How does Industry 4.0 impact on decision-making tasks? In *2018 IEEE Technology and Engineering Management Conference (TEMSCON)* (pp. 1–6). IEEE. https://doi.org/10.1109/TEMSCON.2018.8488409

Apt, W., Bovenschulte, M., Priesack, K., Weiß, C., & Hartmann, E. A. (2018). *Einsatz von digitalen Assistenzsystemen im Betrieb No. 502 in Forschungsbericht*. Berlin: Institut für Innovation und Technik.

Backhaus, N. (2019). Kontextsensitive Assistenzsysteme und Überwachung am Arbeitsplatz: Ein meta-analytisches Review zur Auswirkung elektronischer Überwachung auf Beschäftigte. *Zeitschrift für Arbeitswissenschaft, 73*(1), 2–22. https://doi.org/10.1007/s41449-018-00140-z

Baethge-Kinsky, V., Marquardsen, K., & Tullius, K. (2018). Perspektiven industrieller Instandhaltungsarbeit. *WSI-Mitteilungen, 71*(3), 174–181. https://doi.org/10.5771/0342-300X-2018-3-174

Baumann, H., & Maschke, M. (2016). Betriebsvereinbarungen 2015 - Verbreitung und Themen. *WSI-Mitteilungen, 69*(3), 223–232. https://doi.org/10.5771/0342-300X-2016-3-223

Beer, J. M., Fisk, A. D., & Rogers, W. A. (2014). Toward a framework for levels of robot autonomy in human-robot interaction. *Journal of Human-Robot Interaction, 3*(2), 74–99. https://doi.org/10.5898/JHRI.3.2.Beer

Beernaert, T. F., Bayrak, A. E., Etman, L., Papalambros, P. Y., et al. (2018) Framing the concept of autonomy in system design. In *DS 92: Proceedings of the DESIGN 2018 15th International Design Conference* (pp. 2821–2832).

Berisha-Gawlowski, A., Caruso, C., & Harteis, C. (2020). The concept of a digital twin and its potential for learning organizations. In D. Ifenthaler, S. Hofhues, M. Egloffstein, & C. Helbig (Eds.), *Digital transformation of learning organizations* (pp. 95–114). Berlin: Springer Nature [S.l.]. https://doi.org/10.1007/978-3-030-55878-9_6

Bertschinger, N., Olbrich, E., Ay, N., & Jost, J. (2008). Autonomy: An information theoretic perspective. *Biosystems, 91*(2), 331–345. https://doi.org/10.1016/j.biosystems.2007.05.018

Bibel, W. (2010). General aspects of intelligent autonomous systems. In *Intelligent autonomous systems* (pp. 5–27). Springer, Berlin. https://doi.org/10.1007/978-3-642-11676-6_2

Bläsing, D., & Bornewasser, M. (2021). Influence of increasing task complexity and use of informational assistance systems on mental workload. *Brain Sciences, 11*(1). https://doi.org/10.3390/brainsci11010102

Böhle, F. (2018). Arbeit als handeln. In F. Böhle, G. G. Voß, & G. Wachtler (Eds.), *Handbuch Arbeitssoziologie* (pp. 171–200). Wiesbaden: Springer Fachmedien Wiesbaden.

Bradtke, E., Melzer, M., Röllmann, L., & Rösler, U. (2016). Psychische Gesundheit in der Arbeitswelt - Tätigkeitsspielraum in der Arbeit. https://doi.org/10.21934/baua:bericht20160713/1a

Breaugh, J. A. (1985). The measurement of work autonomy. *Human Relations, 38*(6), 551–570. https://doi.org/10.1177/001872678503800604

Breaugh, J. A. (1999). Further investigation of the work autonomy scales: Two studies. *Journal of Business and Psychology, 13*(3), 357–373. https://doi.org/10.1023/A:1022926416628

Buchholz, V., & Kopp, S. (2021). Towards adaptive worker assistance in monitoring tasks. In *Proceedings of the 2021 IEEE International Conference on Human-Machine Systems (ICHMS)*, IEEE.

Bunt, A., Conati, C., & McGrenere, J. (2004). What role can adaptive support play in an adaptable system? In N. Jardim Nunes & C. Rich (Eds.), *IUI 04* (p. 117). New York: ACM. https://doi.org/10.1145/964442.964465

Burggräf, P., Dannapfel, M., Adlon, T., & Kasalo, M. (2021). Adaptivity and adaptability as design parameters of cognitive worker assistance for enabling agile assembly systems. *Procedia CIRP, 97*, 224–229. https://doi.org/10.1016/j.procir.2020.05.229

Burrell, J. (2016). How the machine 'thinks': Understanding opacity in machine learning algorithms. *Big Data & Society, 3*(1), 205395171562251. https://doi.org/10.1177/2053951715622512

Butollo, F., Jürgens, U., & Krzywdzinski, M. (2018). Von Lean Production zur Industrie 4.0. Mehr Autonomie für die Beschäftigten? *Arbeits-und Industriesoziologische Studien, 11*(2), 75–90.

Choi, S., Leiter, J., & Tomaskovic-Devey, D. (2008). Contingent autonomy: Technology, bureaucracy, and relative power in the labor process. *Work and Occupations, 35*(4), 422–455. https://doi.org/10.1177/0730888408326766

Coyle-Shapiro, J. A. M., & Shore, L. M. (2007). The employee-organization relationship: Where do we go from here? *Human Resource Management Review, 17*(2), 166–179. https://doi.org/10.1016/j.hrmr.2007.03.008

Cropanzano, R., & Mitchell, M. S. (2005). Social exchange theory: An interdisciplinary review. *Journal of Management, 31*(6), 874–900. https://doi.org/10.1177/0149206305279602

Deci, E. L., Olafsen, A. H., & Ryan, R. M. (2017). Self-determination theory in work organizations: The state of a science. *Annual Review of Organizational Psychology and Organizational Behavior, 4*(1), 19–43. https://doi.org/10.1146/annurev-orgpsych-032516-113108

Delfanti, A. (2021). Machinic dispossession and augmented despotism: Digital work in an Amazon warehouse. *New Media & Society, 23*(1), 39–55. https://doi.org/10.1177/1461444819891613

Demerouti, E., & Nachreiner, F. (2019). Zum Arbeitsanforderungen-Arbeitsressourcen-Modell von Burnout und Arbeitsengagement - Stand der Forschung. *Zeitschrift für Arbeitswissenschaft, 73*(2), 119–130. https://doi.org/10.1007/s41449-018-0100-4

Demerouti, E., Bakker, Arnold, & B, Nachreiner F, Schaufeli, Wilmar, B. (2001). The job demands-resources model of burnout. *Journal of Applied Psychology, 86*(3), 499–512. https://doi.org/10.1037/0021-9010.86.3.499

Dourish, P. (2016). Algorithms and their others: Algorithmic culture in context. *Big Data & Society, 3*(2), 205395171666512. https://doi.org/10.1177/2053951716665128

Durkheim, É. (1984). *Erziehung, Moral und Gesellschaft: Vorlesung an der Sorbonne 1902/1903, Suhrkamp Taschenbuch Wissenschaft,* (7th ed., Vol. 487). Frankfurt a. M.: Suhrkamp Verlag.

Elliott, C. S., & Long, G. (2016). Manufacturing rate busters: Computer control and social relations in the labour process. *Work, Employment & Society: A Journal of the British Sociological Association, 30*(1), 135–151. https://doi.org/10.1177/0950017014564601

Endsley, M. R. (2017). From here to autonomy. *Human Factors, 59*(1), 5–27. https://doi.org/10.1177/0018720816681350

Engels, G. (2020). Der digitale Fußabdruck, Schatten oder Zwilling von Maschinen und Menschen. *Gruppe Interaktion Organisation Zeitschrift für Angewandte Organisationspsychologie (GIO), 51*(3), 363–370. https://doi.org/10.1007/s11612-020-00527-9

Eramo, R., Bordeleau, F., Combemale, B., van Den Brand, M., Wimmer, M., & Wortmann, A. (2021). Conceptualizing digital twins. IEEE Software.

Evers, M., Krzywdzinski, M., & Pfeiffer, S. (2018). Designing wearables for use in the workplace: The role of solution developers.

Fischer, H., Engler, M., & Sauer, S. (2018). A human-centered perspective on software quality: Acceptance criteria for work 4.0. In A. Marcus & W. Wang (Eds.), *Design, user experience, and usability: Theory and practice.* Lecture notes in computer science (pp. 570–583). Cham: Springer International Publishing.

Fletcher, S. R., Johnson, T., Adlon, T., Larreina, J., Casla, P., Parigot, L., Alfaro, P. J., & Del Otero, M. M. (2020). Adaptive automation assembly: Identifying system requirements for technical efficiency and worker satisfaction. *Computers & Industrial Engineering, 139,* 105772. https://doi.org/10.1016/j.cie.2019.03.036

Friedman, A. (1977). Responsible autonomy versus direct control over the labour process. *Capital & Class, 1*(1), 43–57. https://doi.org/10.1177/030981687700100104

Gensler, E., & Abendroth, A. K. (2021). Verstärkt algorithmische Arbeitssteuerung Ungleichheiten in Arbeitsautonomie? Eine empirische Untersuchung von Beschäftigten in großen deutschen Arbeitsorganisationen. *Soziale Welt Sonderband, 72*(4), 514–550. https://doi.org/10.5771/0038-6073-2021-4-514

Gil, M., Albert, M., Fons, J., & Pelechano, V. (2019). Designing human-in-the-loop autonomous cyber-physical systems. *International Journal of Human-Computer Studies, 130,* 21–39. https://doi.org/10.1016/j.ijhcs.2019.04.006

Gottschalk-Mazouz, N. (2008). Autonomie und die Autonomie autonomer technischer Systeme. In *XXI. Deutscher Kongress für Philosophie* (Vol. 30). Lebenswelt und Wissenschaft.

Graessler, I., & Poehler, A. (2017). Integration of a digital twin as human representation in a scheduling procedure of a cyber-physical production system. In *Proceedings of the 2017 IEEE IEEM,* IEEE.

Graessler, I., & Poehler, A. (2018). Intelligent control of an assembly station by integration of a digital twin for employees into the decentralized control system. *Procedia Manufacturing, 24*, 185–189. https://doi.org/10.1016/j.promfg.2018.06.041

Gransche, B., Shala, E., Hubig, C., Alpsancar, S., & Harrach, S. (2014). Wandel von Autonomie und Kontrolle durch neue Mensch-Technik-Interaktionen. Grundsatzfragen autonomieorientierter Mensch-Technik-Verhältnisse Stuttgart.

Gräßler, I., Roesmann, D., & Pottebaum, J. (2020). Traceable learning effects by use of digital adaptive assistance in production. *Procedia Manufacturing, 45*, 479–484. https://doi.org/10.1016/j.promfg.2020.04.058

Green, F. (2008). Leeway for the loyal: A model of employee discretion. *British Journal of Industrial Relations, 46*(1), 1–32. https://doi.org/10.1111/j.1467-8543.2007.00666.x

Haase, T., Termath, W., Berndt, D., & Dick, M. (2020). Assistive technologies - companion or controller - appropriation instead of instruction. *Journal of Systemics, Cybernetics and Informatics, 18*(7), 13–18.

Hacker, W., & Sachse, P. (2014). *Allgemeine Arbeitspsychologie: Psychische Regulation von Tätigkeiten* (3rd ed.). Göttingen: Hogrefe.

Heinz, M., & Röcker, C. (2018). Feedback presentation for workers in industrial environments - Challenges and opportunities. In A. Holzinger, P. Kieseberg, A. M. Tjoa, & E. Weippl (Eds.), *Machine learning and knowledge extraction*. Lecture Notes in Computer Science (Vol. 11015, pp. 248–261). Cham: Springer International Publishing. https://doi.org/10.1007/978-3-319-99740-7_17

Humphrey, S. E., Nahrgang, J. D., & Morgeson, F. P. (2007). Integrating motivational, social, and contextual work design features: A meta-analytic summary and theoretical extension of the work design literature. *The Journal of Applied Psychology, 92*(5), 1332–1356. https://doi.org/10.1037/0021-9010.92.5.1332

Ittermann, P., & Niehaus, J. (2018). Industrie 4.0 und Wandel von Industriearbeit - revisited. Forschungsstand und Trendbestimmungen. In H. Hirsch-Kreinsen, P. Ittermann, & J. Niehaus (Eds.), *Digitalisierung industrieller Arbeit* (pp. 33–60). Nomos Verlagsgesellschaft mbH & Co. KG. https://doi.org/10.5771/9783845283340-32

Josifovska, K., Yigitbas, E., & Engels, G. (2019). A digital twin-based multi-modal UI adaptation framework for assistance systems in Industry 4.0. In *Human-computer interaction. Design practice in contemporary societies (HCII 2019)*. Lecture notes in computer science. Cham: Springer.

Kalleberg, A. L., & Vaisey, S. (2005). Pathways to a good job: Perceived work quality among the machinists in North America. *British Journal of Industrial Relations, 43*(3), 431–454. https://doi.org/10.1111/j.1467-8543.2005.00363.x

Kalleberg, A. L., Nesheim, T., & Olsen, K. M. (2009). Is participation good or bad for workers? Effects of autonomy, consultation and teamwork on stress among workers in Norway. *Acta Sociologica, 52*(2), 99–116. https://doi.org/10.1177/0001699309103999

Karasek, R. A. (1979). Job demands, job decision latitude, and mental strain: Implications for job redesign. *Administrative Science Quarterly, 24*(2), 285. https://doi.org/10.2307/2392498

Kellogg, K. C., Valentine, M. A., & Christin, A. (2020). Algorithms at work: The new contested terrain of control. *Academy of Management Annals, 14*(1), 366–410. https://doi.org/10.5465/annals.2018.0174

Kubicek, B., Korunka, C., & Tement, S. (2014). Too much job control? Two studies on curvilinear relations between job control and eldercare workers' well-being. *International Journal of Nursing Studies, 51*(12), 1644–1653. https://doi.org/10.1016/j.ijnurstu.2014.05.005

Kubicek, B., Paškvan, M., & Bunner, J. (2017). The bright and dark sides of job autonomy. In Korunka, C., & Kubicek, B. (Eds.), Job demands in a changing world of work (pp. 45–63). Cham: Springer International Publishing. https://doi.org/10.1007/978-3-319-54678-0_4

Kuhlmann, M., Splett, B., & Wiegrefe, S. (2018) Montagearbeit 4.0? Eine Fallstudie zu Arbeitswirkungen und Gestaltungsperspektiven digitaler Werkerführung. *WSI-Mitteilungen, 71*(3), 182–188. https://doi.org/10.5771/0342-300X-2018-3-182

Lammi, I. J. (2021). Automating to control: The unexpected consequences of modern automated work delivery in practice. *Organization, 28*(1), 115–131. https://doi.org/10.1177/1350508420968179

Langfred, C. W. (2005). Autonomy and performance in teams: The multilevel moderating effect of task interdependence. *Journal of Management, 31*(4), 513–529. https://doi.org/10.1177/0149206304272190

Langfred, C. W., & Moye, N. A. (2004). Effects of task autonomy on performance: An extended model considering motivational, informational, and structural mechanisms. *The Journal of Applied Psychology, 89*(6), 934–945. https://doi.org/10.1037/0021-9010.89.6.934

Loch, F., Czerniak, J., Villani, V., Sabattini, L., Fantuzzi, C., Mertens, A., & Vogel-Heuser, B. (2018). An adaptive speech interface for assistance in maintenance and changeover procedures. In M. Kurosu (Ed.), *Human-computer interaction. Interaction technologies*. Lecture Notes in Computer Science (Vol. 10903, pp. 152–163). Cham: Springer International Publishing. https://doi.org/10.1007/978-3-319-91250-9_12

Lyons, J. B., Sycara, K., Lewis, M., & Capiola, A. (2021). Human-autonomy teaming: Definitions, debates, and directions. *Frontiers in Psychology, 12*, 589585. https://doi.org/10.3389/fpsyg.2021.589585

Mark, B. G., Rauch, E., & Matt, D. T. (2021). Worker assistance systems in manufacturing: A review of the state of the art and future directions. *Journal of Manufacturing Systems, 59*, 228–250. https://doi.org/10.1016/j.jmsy.2021.02.017

Meijerink, J., & Bondarouk, T. (2021). The duality of algorithmic management: Toward a research agenda on HRM algorithms, autonomy and value creation. *Human Resource Management Review*, 100876. https://doi.org/10.1016/j.hrmr.2021.100876

Morana, S., Pfeiffer, J., & Adam, M. T. P. (2020). User assistance for intelligent systems. *Business & Information Systems Engineering, 62*(3), 189–192. https://doi.org/10.1007/s12599-020-00640-5

Morgeson, F. P., & Humphrey, S. E. (2006). The work design questionnaire (WDQ): Developing and validating a comprehensive measure for assessing job design and the nature of work. *The Journal of Applied Psychology, 91*(6), 1321–1339. https://doi.org/10.1037/0021-9010.91.6.1321

Muecke, S., & Iseke, A. (2019). How does job autonomy influence job performance? A meta-analytic test of theoretical mechanisms. *Academy of Management Proceedings, 1*, 14632. https://doi.org/10.5465/AMBPP.2019.145

Mühge, G. (2018). Einzug der Rationalität in die Organisation? Digitale Systeme der Entscheidungsunterstützung in der Produktion. *WSI-Mitteilungen, 71*(3), 189–195. https://doi.org/10.5771/0342-300X-2018-3-189

Murphy, R. R. (2000). *Introduction to AI robotics*. MIT Press.

Oestreich, H., Töniges, T., Wojtynek, M., & Wrede, S. (2019). Interactive learning of assembly processes using digital assistance. *Procedia Manufacturing, 31*, 14–19. https://doi.org/10.1016/j.promfg.2019.03.003

Oestreich, H., Wrede, S., & Wrede, B. (2020). Learning and performing assembly processes. In: Makedon, F. (Eds.), *Proceedings of the 13th ACM International Conference on PErvasive Technologies Related to Assistive Environments* (pp. 1–8). Association for Computing Machinery, [S.l.]. https://doi.org/10.1145/3389189.3397977

Oestreich, H., Da Silva Bröker, Y., & Wrede, S. (2021). An adaptive workflow architecture for digital assistance systems. In *The 14th PErvasive Technologies Related to Assistive Environments Conference* (pp. 177–184). New York, NY, USA: ACM. https://doi.org/10.1145/3453892.3458046

Oestreich, H., Heinz, M., Sehr, P., & Wrede, S. (2022). Human-centered adaptive assistance systems for the shop floor. In S. Büttner, C. Röcker (Ed.), *Human technology interaction - Shaping the future of industrial user interfaces*. Berlin: Springer International Publishing (in press).

O'Neill, T., McNeese, N., Barron, A., & Schelble, B. (2020) Human-autonomy teaming: A review and analysis of the empirical literature. *Human Factors*, 18720820960865. https://doi.org/10.1177/0018720820960865

Orlikowski, W. J. (2000). Using technology and constituting structures: A practice lens for studying technology in organizations. *Organization Science, 11*(4), 404–428. https://doi.org/10.1287/orsc. 11.4.404.14600

Parker, S. K., & Grote, G. (2020). Automation, algorithms, and beyond: Why work design matters more than ever in a digital world. *Applied Psychology: An International Review*. https://doi.org/ 10.1111/apps.12241

Peterson, H. (2019). How Amazon tracks and scores delivery drivers on the road. Business Insider.

Pongratz, H. J., & Voß, G. G. (2003). From employee to 'entreployee'. *Concepts and Transformation, 8*(3), 239–254. https://doi.org/10.1075/cat.8.3.04pon

Rohlfing, K. J., Cimiano, P., Scharlau, I., Matzner, T., Buhl, H. M., Buschmeier, H., Esposito, E., Grimminger, A., Hammer, B., Häb-Umbach, R., Horwath, I., Hüllermeier, E., Kern, F., Kopp, S., Thommes, K., Ngomo, A. C. N., Schulte, C., Wachsmuth, H., Wagner, P., & Wrede, B. (2020). Explanation as a social practice: Toward a conceptual framework for the social design of AI systems. *IEEE Transactions on Cognitive and Developmental Systems*, 1–1. https://doi.org/10. 1109/TCDS.2020.3044366

Rosen, R., von Wichert, G., Lo, G., & Bettenhausen, K. D. (2015). About the importance of autonomy and digital twins for the future of manufacturing. *IFAC-PapersOnLine, 48*(3), 567–572. https://doi.org/10.1016/j.ifacol.2015.06.141

Sankaran, S., Zhang, C., Aarts, H., & Markopoulos, P. (2021). Exploring peoples' perception of autonomy and reactance in everyday AI interactions. *Frontiers in Psychology, 12*. https://doi.org/ 10.3389/fpsyg.2021.713074

Schildt, H. (2017). Big data and organizational design - The brave new world of algorithmic management and computer augmented transparency. *Innovation, 19*(1), 23–30. https://doi.org/10. 1080/14479338.2016.1252043

Schilling, M., Kopp, S., Wachsmuth, S., Wrede, B., Ritter, H., Brox, T., Nebel, B., & Burgard, W. (2016). Towards a multidimensional perspective on shared autonomy.

Schwarz, J., Fuchs, S., & Flemisch, F. (2014). Towards a more holistic view on user state assessment in adaptive human-computer interaction. In *2014 IEEE International Conference on Systems, Man, and Cybernetics (SMC)* (pp. 1228–1234). IEEE. https://doi.org/10.1109/SMC.2014. 6974082

Schweden, F., Kästner, T., & Rau, R. (2019). Erleben von Tätigkeitsspielraum. *Zeitschrift für Arbeits- und Organisationspsychologie A&O, 63*(2), 59–70. https://doi.org/10.1026/0932-4089/ a000280

Selke, S., & Biniok, P. (2015). Assistenzensembles in der Gesellschaft von morgen. In *AAL-Kongress 2015*. Frankfurt/Main, Deutschland: VDE Verlag.

Simmler, M., & Frischknecht, R. (2021). A taxonomy of human-machine collaboration: Capturing automation and technical autonomy. *AI & Society, 36*(1), 239–250. https://doi.org/10.1007/ s00146-020-01004-z

de Spiegelaere, S., van Gyes, G., de Witte, H., Niesen, W., & van Hootegem, G. (2014). On the relation of job insecurity, job autonomy, innovative work behaviour and the mediating effect of work engagement. *Creativity and Innovation Management, 23*(3), 318–330. https://doi.org/10. 1111/caim.12079

de Spiegelaere, S., van Gyes, G., & van Hootegem, G. (2016). Not all autonomy is the same. Different dimensions of job autonomy and their relation to work engagement & innovative work behavior. *Human Factors and Ergonomics in Manufacturing & Service Industries, 26*(4), 515–527. https:// doi.org/10.1002/hfm.20666

Stanton, J. M., & Barnes Farrell, J. L. (1996). Effects of electronic performance monitoring on personal control, task satisfaction, and task performance. *The Journal of Applied Psychology, 81*(6), 738–745.

Strauch, B. (2018). Ironies of automation: Still unresolved after all these years. *IEEE Transactions on Human-Machine Systems, 48*(5), 419–433. https://doi.org/10.1109/THMS.2017.2732506

Tomczak, D. L., Lanzo, L. A., & Aguinis, H. (2018). Evidence-based recommendations for employee performance monitoring. *Business Horizons, 61*(2), 251–259. https://doi.org/10.1016/j.bushor.2017.11.006

Töniges T., Ötting, S. K., Wrede, B., Maier, G. W., & Sagerer, G. (2017). An emerging decision authority. In *Cyber-physical systems* (pp. 419–430). Amsterdam: Elsevier. https://doi.org/10.1016/B978-0-12-803801-7.00026-2

Ullrich, C. (2016). Rules for adaptive learning and assistance on the shop floor. In *13th International Conference on Cognition and Exploratory Learning in Digital Age* (pp. 261–268). IADIS Press.

Varadinek, B., Indenhuck, M., & Surowiecki, E. (2018) Rechtliche Anforderungen an den Datenschutz bei adaptiven Arbeitsassistenzsystemen. https://doi.org/10.21934/baua:bericht20180820

Weitz, K., Schiller, D., Schlagowski, R., Huber, T., & André, E. (2021). Let me explain!: Exploring the potential of virtual agents in explainable AI interaction design. *Journal on Multimodal User Interfaces, 15*(2), 87–98.

Wendemuth, A., Böck, R., Nürnberger, A., Al-Hamadi, A., Brechmann, A., & Ohl, F. W. (2018). Intention-based anticipatory interactive systems. In *2018 IEEE International Conference on Systems, Man, and Cybernetics (SMC)* (pp. 2583–2588). https://doi.org/10.1109/SMC.2018.00442

Wheatley, D. (2017). Autonomy in paid work and employee subjective well-being. *Work and Occupations, 44*(3), 296–328. https://doi.org/10.1177/0730888417697232

Wood, A. J. (2021). Algorithmic management: Consequences for work organisation and working conditions.

Woolf, B. P. (2009). *Building intelligent interactive tutors: Student-centered strategies for revolutionizing e-learning.* Amsterdam and Boston: Morgan Kaufmann Publishers / Elsevier.

Yigitbas, E., Jovanovikj, I., Biermeier, K., Sauer, S., & Engels, G. (2020). Integrated model-driven development of self-adaptive user interfaces. *Software and Systems Modeling.* https://doi.org/10.1007/s10270-020-00777-7

Zibuschka, J., Ruff, C., Horch, A., & Roßnagel, H. (2020). A human digital twin as building block of open identity management for the Internet of Things, 1617–5468. https://doi.org/10.18420/ois2020_11

Individual Assembly Guidance

Alexander Pöhler and Iris Gräßler

Abstract In this chapter, an approach to assisting manual assembly processes individually is presented. The objective is to reduce process times by giving individual instructions to the assemblers. In manual assembly, worker assistance systems are used for worker guidance and the simultaneous quality assurance of the assembly steps in real time. The assistance system guides the workers step by step through the assembly process and simultaneously ensures that no errors occur. In practice, it has been shown that the use of assistance systems increases the quality of the assembly process but also increases process times. The approach shown in this chapter overcomes this loss in efficiency by assisting workers individually. The assistance system learns from the process executions of the employees through the digital twins of the employees. The assistance system adapts to the behaviour of the employees and highlights process steps in which mistakes might occur. Thereby, the number of interactions with the assistance system is reduced and more specific assistance is given.

Keywords Industry 4.0 · Assistance system · Assembly guidance

1 Introduction

In recent years, the topic of networked industrial machines for manufacturing companies has increasingly become the focus of serious research and development. The term 'Industry 4.0' is used to summarise concepts that are intended to produce individual products automatically (Kagermann et al., 2011). A major guideline for such an automatised individual production system is that the schedule of a workpiece should not be managed by a higher-level control system; instead, it should be carried

A. Pöhler (✉) · I. Gräßler
Heinz Nixdorf Institute, Paderborn University, Fürstenallee 11, 33102 Paderborn, Germany
e-mail: alexander.poehler@hni.upb.de

I. Gräßler
e-mail: iris.graessler@hni.upb.de

© The Author(s), under exclusive license to Springer Nature Switzerland AG 2023   171
I. Gräßler et al. (eds.), *The Digital Twin of Humans*,
https://doi.org/10.1007/978-3-031-26104-6_8

along with the object to be processed. This can, for example, be achieved using a decentralised production control system. The machines used in such an environment are implemented as intelligent agents that can independently carry out the production plan. In this process, the workpiece carries its assembly sequence plan with it from workstation to workstation in its so-called 'digital product memory' (Kagermann et al., 2011). This is one way to carry out production planning and control for small-batch-size complex products. In assembly processes, however, workers are confronted with a great variety of products that need to be assembled. This increases the complexity of the assembly task, and some kind of support is needed in order to maintain short processing times. In order to implement such an assistance system for a great variety of products, in the past, assembly instructions for all variants needed to be available. The assistance system was then mainly used to show the worker how to assemble the product, but always in the same sequence. Thereby, the complexity of the assembly processes of a variety of products can be handled. However, the processing times increase due to complex assembly tasks and also the operation of the assistance system.

The assembly processes of complex products with small batch sizes can differ greatly in terms of the duration of processing and the way that the assembly process is carried out. A worker assistance system can support workers in managing this complex assembly task. The assistance system can be used to display digital instructions to workers so that they can successfully perform their assembly tasks. The information presented by such a worker assistance system should be derived from the product's assembly flowchart, so that a new assembly flowchart does not have to be created from scratch when a product variant is created. If the assembly flowchart is designed such that it also contains work instructions for manual assembly in addition to the assembly sequence, the information that is important for workers can be derived directly from this flowchart and interpreted by an assistance system.

To support the worker with assistance information, an assistance system must have knowledge about the operation that is currently being performed by the worker. Accordingly, an internal representation of the activity that requires assistance is necessary. In general, assembly instructions are mostly provided in paper form, but these instructions are inflexible with respect to single-item production and small-batch production with frequent changes or variations in product families. To remain competitive, more flexibility must be brought to manual workstations. In order to respond efficiently to variations in production, simple definitions of assembly flowcharts and the associated work instructions therefore must be implemented.

The assistance system must always know which work process the worker is currently performing in order to display assembly instructions and additional information. In this approach, the recognition of the current work process is performed non-invasively using a camera-based recording of the work area. The main indicators of the current working process for the assistance system are the positions of the hands and their movement and the status of the assembled object itself. By analysing the recorded images, the objects of the assembly can be derived as features, which can then be tracked by the assistance system. Together with the hand movements, which indicate which parts were grabbed, this information can be used to perform an automatic determination of whether an assembly step is finished or not.

With the knowledge of the assembly activity and the observations of the recorded events in the manual assembly process, assistance information can be given to the worker as support. Accordingly, a visualisation system that extracts and displays this information from the assembly flowchart is required. In general, assembly flowcharts describe the sequence of assembly activities that workers have to carry out in industrial production.

The correct assembly of products is carried out on the basis of sequence instructions, which are also known as assembly sequence plans. For the realisation of a worker assistance system, it is necessary to define a technical representation that depicts the assembly process holistically or to a relevant extent, so that it can provide adequate assistance to the worker. This representation can then be used to automatically interpret and further process the underlying flowchart to create an assembly flowchart. The representation thereby consists of data that describe how the employee executes tasks. For the assistance system, processing times and errors during assembly are used for evaluation.

The individualisation of the assembly instructions is meant to create a user-centred execution of the assembly process. It actually involves the implementation of the digital twins of employees in this context. The individualisation is carried out by using the process times of the employee, his preferences, and all other information that can be used to support the assembly instructions. The usage of these data in this scenario relies on the concept of the digital twins of employees.

In this chapter, the development of a human-centred assistance system for assembly is shown. The novelty of this approach is that it adapts assembly guidance for different users. Additionally, this approach includes a method for automatically deriving the assembly instructions and assembly steps. The approach of the assistance system is to take a full set of assembly steps and then focus on the relevant steps. Therefore, at first, a full set of assembly steps needs to be available to the assistance system. Then, the actual execution of the assembly process is analysed. This analysis is done individually, which means that for every worker using the assistance system, their execution, mistakes, and the duration of each step are tracked. Based on this analysis, the assembly assistance process is modified. Thereby, workers receive their personal assembly instructions, which are based on past mistakes and steps that have taken them a long time, i.e., longer than the average execution time. In order to be able to track the assembly progress, the work steps need to be tracked automatically. The novelty of this approach mostly involves tracking the working steps and adapting the assembly instructions based on the analysis of these steps.

In the following section, the state of the art for assembly guidance systems is described. Then, the automatic determination of the work steps of the developed assistance system is presented. Next, the implementation of the assistance system is described in Sect. 4, and the results of the usage of the assistance system are shown and discussed. The chapter concludes with Sect. 5.

2   State of the Art

According to Dangelmaier (2009), the term 'assembly process' refers to the grouping of operations that create a product with a new identity from several components of different identities. This process is also referred to as an assembly task in the following. The essential characteristic of an assembly task is that an operation can only be completed when all the necessary components are available. This results in the need for all required components to be ready at the assembly site (Padoy et al., 2009). Assembly operations are described using an assembly process structure. Within this description of the assembly task, the dependencies of the assembly activities to be performed are also included. In general, the structure of the assembly sequence can be divided into the task breakdown and the sequence breakdown. The task breakdown reflects the subdivision of the assembly task into possible assembly activities. Partial tasks or subtasks are defined. Thus, a purpose-oriented planning process is made possible. This division of assembly activities is carried out at varying levels of detail depending on the production process. In large-scale production and mass production, the task breakdown is carried out at the level of work operations (Glonegger et al., 2013). A work operation is generally described by several assembly activities. Examples of these assembly activities are activities such as gripping or turning. The actions to be performed during an assembly activity are also defined in Verband deutscher Ingenieure (1990). The higher demand for individualised products with special features or individual components has resulted in a variety of challenges in production. Smaller batch sizes, complex products, and high quality requirements have contributed to a noticeable increase in performance pressure for all involved. In manual assembly, this leads to decreasing process quality and increasing process times. In order to meet the associated challenges, assistance systems adapted to the assembly process can be used to support workers (Zäh & Reinhart, 2014).

Employees in variant or multi-product assembly must increasingly perform informational work, as they have to make selection decisions based on different products, tools, or aids (Bornewasser et al., 2018). Technical support is needed to control these dynamic-value-creation processes, and assistance systems are often used. They provide context-relevant information or communication options. The employees are supported in carrying out the assembly process and the error potential is reduced. This is accompanied by an improvement in the process stability, as well as lower reject rates and quality defects (Hinrichsen et al., 2018). Worker assistance systems are divided into three categories (Apt et al., 2016):

1. Assistance systems make existing analogue knowledge available in a digital form or provide work instructions via a simple display (e.g., digital manuals, learning videos, assembly or maintenance instructions, quality instructions, safety instructions, process knowledge, qualification management, classical knowledge management systems).

2. Adaptive assistance systems allow the sensory recording of the work processes, and the context in which they take place, of the respective user, and they can be individually adapted by the employees (e.g., worker guidance, pick-by-light, context-sensitive information provision, adaptation of language or user interfaces).

3. Tutorial assistance systems are adaptive assistance systems that support a working environment conducive to learning in the process of work ('learning tools'). They enable the individual provision of situational, relevant knowledge (e.g., education and training systems, portable training systems, portable knowledge and learning platforms, augmented reality-based support systems for technical service).

Display technology allows the presentation of an almost unlimited amount of information. Beyond pure decision support (e.g., work instructions), additional documentation, communication procedures, or hazard warnings can be realised. Screens can be stationary (at a workplace) and mobile. If they also have an input function, they represent a multimodal interface that enables the worker to interact with the company's ERP system (Schreiber, 2017).

The pick-by-light system supports the operator during material removal. A light signal indicates to the worker which box the correct material should be picked up from. The light pulse eliminates the time needed for reading about, searching for, identifying, and checking the correct material. The assembly time can be reduced accordingly with the help of this system (Arnold & Furmans, 2009).

A survey has shown that one third of the German industrial companies surveyed use worker assistance systems (Klapper et al., 2020). The main area of application of these systems is small batch assembly (Klapper et al., 2020; Bannat, 2014). The most important functions of the assistance systems are support of the work task, quality assurance, and process management (Blumberg & Kauffeld, 2020). According to the survey, assistance in assembly is mostly realised through visualisation.

Crucial for the acceptance of digital assistance systems are aspects of work and motivation psychology, which must be addressed within the framework of a human-centred design process with the involvement of the future users of these systems. For the design of work and everyday objects, uniform standards and methods are required; they are described in a basic standard (DIN Deutsches Institut für Normung e. V. Normenausschuss Ergonomie, 2002). The aim of this series of standards is to prevent damage to health and to optimise the user-friendliness of processes and dialogues involving machines and computers. In such a socio-technical approach, the levels of people, organisation, and technology are considered in an integrated way, and they are the basis for a positive 'user experience' and a high degree of 'usability' in the use of digital assistance systems. Typical user requirements for the design of digital assistance systems in the context of assembly are, for example, a general improvement of work processes, an ergonomically improved presentation of relevant information at the workplace, and physical and mental relief, as well as the low-effort successful implementation of an assistance system (Hinrichsen et al., 2018). It is critical to the development of digital assistance systems to take into account the needs of the user, as well as the cognitive and emotional states and work processes of the users. Therefore, providing assembly instructions individually is the key not only to decreasing process times, but also to the acceptance of assistance systems.

In the following, an approach to providing assembly guidance individually is presented. The approach is implemented in a laboratory. This will improve the applicability of this concept and show its relevance to industry. The underlying concept of individual assembly guidance represents a novel approach. Therefore, further studies observing and evaluating this concept still need to be carried out. Its implementation, however, has a significant scientific impact and will open up the possibility of further evaluating this concept. In the following section, the basic principles of the developed assistance system are described. Its main task is to determine the current work steps. This process becomes especially relevant for individual assembly guidance, because it gives workers the ability to change the assembly sequence, for example, by skipping or swapping work steps. The assistance system needs to determine whether this change is acceptable or unacceptable.

3 Automatic Determination of Work Steps

Assistance systems for assembly provide instructions regarding the current process step. Through inputs into the assistance system, the current process step and the status of the assembly station and the worker are determined. Inputs are actions that workers perform during their assembly activities. With the help of the assembly schedule, these activities can then be used to draw conclusions about the worker's actions and the associated work process. In the following, the determination of work steps for this assistance system is described.

During the assembly activity, a worker carries out different actions, which are called worker actions. This term is used to describe those activities that the worker performs at the moment of observation. The term 'observation' refers to the sensory acquisition of these worker actions. Furthermore, worker actions may also involve physical actions, such as the movement of the worker between different locations. Accordingly, non-invasive worker action detection techniques and methods that do not require stationary sensors are of interest. The assembly system in which the assistance system is implemented is shown in Fig. 1. The assembly station is part of a production laboratory. The laboratory is designed as a research infrastructure for Industry 4.0 settings. Current configurations are focused on the flexible reconfiguration of production control systems through the decentralisation of decision-making and execution management. The laboratory's main purpose is the evaluation of concepts developed in foundational or applied research. The manufacturing is carried out by three different machines, which are connected via a material flow system. After the manufacturing processes are complete, parts are delivered to the assembly station. The assembly station was built with an aluminium profile and, in addition to the Raspberry Pi control system, essentially consists of a touch screen, depth monitor, and self-developed pick-by-light system with a material feed and the corresponding

Fig. 1 Assembly station with which the assistance system is evaluated

tools for assembly. The developed assembly guidance process is an essential part of the program for controlling the assembly station.

The control system of the assembly system is presented by Gräßler and Pöhler (Graessler & Poehler, 2018). In order to provide the right assistance at the right time, the assistance system must know which components have been removed from which boxes on the work surface. To do this, these boxes must first be found. In order to be able to recognise the positions of the boxes, the image from the camera mounted on top of the assembly station is subjected to colour segmentation. Thus, the image-based detection of these boxes can be performed using the colour information of the camera pictures.

One important assembly activity is accessing components stored in boxes. To perform this activity, it is necessary to identify not only the box but also the hand and its movement, and the system should determine if a part was grabbed from the box. By determining the hand position of the worker, box access can be inferred. To access a box, the worker's hand has to perform a specific motion sequence: The hand has to move forward, downward, upward, and finally backward. Whether the worker has performed this movement is determined by a state machine. Over the scanning time, the hand position data are transmitted to the state machine. If the hand moves out of the work area towards a box, a gripping process begins. With knowledge about the locations of the boxes on the work surface, in combination with the hand position, the system can determine over or in which box the hand is located. Since the system has a knowledge base that contains information about which component is in which box, the type of component that is gripped can be inferred from box access.

The storage of the assembly is transmitted as an event with the storage location; this is analogous to component removal. When a previously stored assembly is subsequently removed, this process can again be transmitted to the assembly schedule assistant as an event. The inputs are actions that workers perform during their assembly activities. With the help of the assembly schedule, these activities can then be used to draw conclusions about the worker's actions and the associated work process. After completing the work step, the worker will grip a new component. Thus, the worker has to reach into a component box. This grip is an event that represents the start of a new work step. The higher-level evaluation of the events in the assembly schedule can be used to determine whether the worker has grabbed the correct component. Provided that the worker grips the correct component, this event also provides the signal that the previous work step is to be considered completed. In the case of an incorrect gripping event, the system treats the current work step as completed only after the correct component has been gripped.

This system allows workers to flexibly adapt their assembly task to their own needs. In the event of possible intermediate steps, they can interrupt their assembly task by placing the assemblies they have just produced into a certain box. The assembly schedule assistant recognises this deposit and saves the contents of the box. The worker can now continue on to another work step. If the worker reaches the point in the assembly schedule where he needs the previously stored assembly again, the worker can retrieve it from the corresponding box. The worker can thus continue the previously paused work process. The assembly schedule assistant will then infer from the grabbed assembly which work step the worker is currently performing. The worker is therefore not required to rigidly follow the assembly schedule. This approach to determining a finished process step is derived from the following processes:

1. Identification of hands

   a. Gripping a component
   b. Gripping a tool
   c. Reaching into a box

2. Event of pick-by-light system
3. Reaching process timer
4. Manual operation via GUI.

After all these processes, an event is triggered, and the system determines whether the work step is finished or not. By measuring the time taken to complete each work step and by noting mistakes in the assembly process through quality control, the system figures out how to adapt the assembly plan. If someone, for example, tries to assemble pieces before the previous tasked is finished, a warning message is given only the first two times that this occurs. If the assembled product makes it through quality control, the assembly plan is adapted accordingly. The assistance system then expects that the worker will want to alter the assembly plan and therefore adapts. In the following section, the implementation of the assistance system is shown, and the individual approach is presented.

4 Implementation

This section presents the implementation of the assistance system described in the previous section. The basis for this investigation is a use case that involves the manual assembly of a product that is new and unknown to the worker. In order to enable a quick training period, an emphasis is placed on the comprehensibility and easy operability of such an assistance system. The worker is supported by an assistance system that provides the necessary information on the work process digitally. First, conventional paper-based work instructions are examined and the underlying information is extracted from them. This information is presented to the worker using a monitor-based GUI, which is shown in Fig. 2.

A digital assembly flowchart is presented; it was implemented using the display of the work instructions in the worker's workspace. In industrial assembly lines,

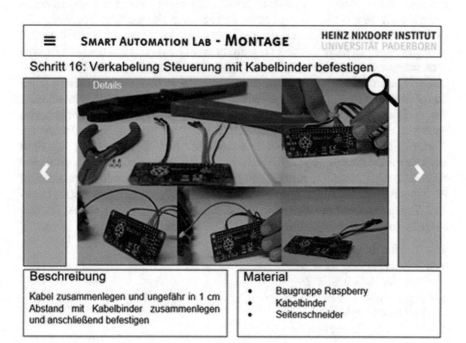

Fig. 2 GUI of the assistance system

there are few methods for displaying work instructions in productive operations. Paper instructions are often used to indicate the necessary steps for the assembly of a product. Workers are instructed on how to perform their task when they are assigned a new work position. Since the assembly sequence generally does not change within the series, the workers do not need any further detailed instructions. They always carry out the same assembly step. In small batch production, product changes within the line may occur more frequently. Accordingly, the worker in this case is mainly presented with assembly flowcharts in paper form that contain and explain the work steps to be performed. Depending on the complexity of the workpiece to be produced, these assembly flowcharts have varying levels of detail. In the case of individual production, each individual workpiece must be accompanied by its own assembly flowchart. Each work step is not carried out very frequently in this case.

The associated work steps are explained in tabular form in the lower part of the work instructions. The sequence of the work steps is determined by the order in which they are written. In addition, a number for each work step is entered in colour in the table; this number can also be found in the photo of the assembly. The number is located at the position where the work steps described in the table are to be performed by the worker. The definition of work instructions and the associated display system for interpretation and visualisation must be flexible enough to accommodate the above information. To achieve this goal, different technical representation methods could be used. In this case, XML is generally used to store the data of the representation. XML makes it possible to define and structure data and data constructs hierarchically in text files. The following example explains the basic structure of the developed assembly flowchart in the XML meta-language, with reference to manual assembly. The assembly flowchart is reproduced in XML notation below. As can be seen, components are defined by naming them. Assemblies are defined by their associated dependencies with respect to the components to be assembled. The required components are named for each assembly within the part description. In addition, the number of parts of each type required for the currently defined assembly is also specified. With this type of definition, each assembly can be a component for another work step in the assembly process. This provides a high degree of flexibility. Each work step must therefore be described only for itself. Interrelationships between the individual work steps are automatically created later via the recursive linking of the component descriptions. If an assembly flowchart has to be extended, the corresponding assemblies have to be changed or added at the appropriate points. The new overall structure of the assembly flowchart then results automatically.

- < ?xml version="1.0" >
- <Product Name="RC-CAR">
- <Part Name="Chassis" />
- <Part Name="Battery" />
- <Part Name="Motor" />
- <Part Name="Electronic" />
- <Part Name="Cables" />
- <Part Name="Body" />

- <Part Name="Rims" />
- <Part Name="Tyre" />.

The next step in the development of an assistance system for manual assembly is the enrichment of the contact analogue display with assistance information, in addition to the work instructions. This assistance information is intended to support the worker in performing a task. To realise this requirement, the concept of the context bubble was extended by a supporting component. The system shows the worker the components he needs for the current work step. This was achieved in the context bubble with the circular arrangement of the corresponding images. One way to support the worker is to help him find the right components. Pick-by-light is a variant in which light signs are used to direct the worker's attention to relevant areas of the workstation. These light signs are usually stationary lamps mounted on shelves, and they are intended to assist the worker in picking components and packing orders. The workstation is meant to be flexible. An alternative to lamps is the active illumination of the components by means of the beamer. The beamer is located above the work surface and can thus be used as a light source. The assistance system was then put in a case. All of the relevant functions were contained in this case. A mini-PC was used to run the application, and cameras and projectors were implemented in order to create the aforementioned functionalities. However, no 3D camera was used, although this was shown in Fig. 1. The small version of the assistance system (in a case) is shown in Fig. 3.

In the following, the individual approach of the assistance system is explained. The normal approach for assembly guidance is to simply show assembly instructions for the step that the user is performing at the moment. Sometimes, some action, such as documenting the assembly progress by taking a picture, is performed. Through the automatic determination of work steps, the assistance system switches automatically between most work steps and does not need to be operated. Only if no determination of whether the current step is complete is possible does the worker have to switch to the next assembly step manually. Current assembly guidance systems function in this way: They show the assembly instructions and normally have some camera-

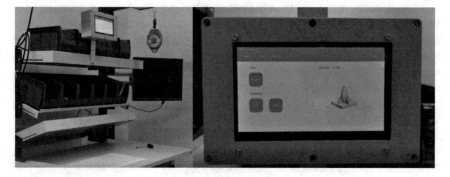

Fig. 3 Small version of the assistance system with its GUI

based quality control tools. The idea of individually assisting assembly is to adapt the assembly instructions to the worker. The implementation of this system is described using the following expansion stages.

1. The easiest way to adapt the assembly instructions is to provide individual hints. This is realised by measuring past mistakes that were made during process execution. For example, a part may have been forgotten by a worker during assembly. An easy way to assist the worker is to highlight this assembly step and therefore remind the worker to assemble this part. For implementation, it is necessary to gather this information through quality control, which is normally carried out at the end of an assembly sequence.
2. The next step is to determine the process times for each assembly step and thereby anticipate how long each assembly step will take. Thereby, it is possible to automatically proceed with assembly instructions without having the worker operate the assistance system. The idea is that after the average process time of the assembly step has passed, the next assembly instructions are shown. Thereby, the automation of process steps for which the completion of the step cannot be identified can be carried out.
3. The third step is to monitor the worker's behaviour. At which steps does he look at the assembly guidance system, and how does he interact with it in order to look something up? The idea is to highlight these steps.
4. Based on the third step, a further development of this idea is to only show the steps at which the worker interacted with the assistance system. This is only possible if the worker has carried out the assembly process for this product several times.
5. The last step is to allow alterations to the assembly sequence. If, for example, a worker wants to assemble the part in another way and he has the experience to do so, the assembly guidance system works mostly as a quality control system that gives hints at certain steps. This way, the worker can adapt the process fully to his own preferences and thereby decrease process times. In order to carry out such an approach, which process steps can be altered must be predefined.

The expansion stages show the basic implementation of individual assembly guidance on different levels. All of these levels have been implemented at the assembly station and functionality tests have been carried out. The assistance system was evaluated for different production orders for which different products needed to be produced in small lot sizes. The assistance system learns quite fast to adapt to a user, and thereby it adapts its assembly plans accordingly. The assistance system was evaluated through the assembly of demonstrators, which were produced in a production laboratory. The demonstrator is a miniature drilling machine; it is shown in Fig. 4.

Different tests were performed on the assistance systems in order to evaluate whether an individual assembly guidance system would be beneficial for workers. Unfortunately, most process steps for assembly are fixed and cannot be changed by workers. Therefore, much of the assembly guidance needs to be shown and cannot be altered. The contribution of individually adapting the assembly sequence and the tips and assembly instructions shown is therefore limited to certain assembly processes. If a company wants to monitor this assembly process, it needs to define

Fig. 4 Demonstrator used
for assembly analysis

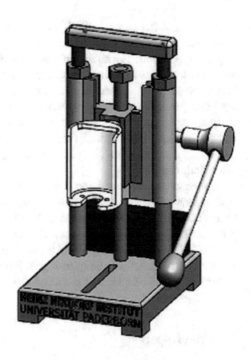

which processes should not be altered by the system. For the miniature drilling machine, only the process steps of the prefabricated parts, which are also assembled at the assembly station, can be altered. A comparison was done between an assembly for which no guidance was given and an assembly for which the assembly guidance system had to be used, and after each process step, the worker had to click to get to the next step. The assembly of the miniature drilling machine has in total 42 process steps.

The operation of the assembly guidance system led to longer process times. The operation of the assembly guidance system led to execution times that were approximately 3–4 s longer per process step. The first step was to implement the automatic determination of process steps. That lead to a direct reduction of process times by 2 s per process step. Then, the individual assembly approach was implemented extension by extension. The implementation was carried out using an experimental setup. Studies that involve measuring times and comparing parameters should follow in order to evaluate this concept. For this exemplary demonstration, it could not be determined if the shortened process times were due to more training and more repetitions of the assembly tasks or due to the implementation of a new extension level of individual assembly guidance. The evaluation is therefore merely based on the feedback of the test persons. The first extension, which consisted of giving hints, was greatly appreciated. Important process steps were highlighted, and therefore the guidance system acted more like a reminder, specifying which steps the assembler should pay more attention to. The automatic onward clicking helped to greatly reduce the process times for these steps, especially if these steps were not highlighted before. Since the assistance system does not need to be operated for these steps, up to 4 s per process step could be saved. The monitoring of the workers' behaviour helped to improve the assembly guidance in such a way that the test persons felt that the assembly guidance system was helpful to them and not merely a control tool. The process times were not greatly reduced by behaviour monitoring, and they were also not reduced by not showing certain process steps, since the test person did not look at these process steps on the assistance system display regardless. The last step, which consisted of individually adapting the assembly sequence, could not be properly tested with the given demonstrator.

The processes executed showed that the assistance system adapts its assembly instructions to the workers. Therefore, the feasibility of this system was shown. The process times were greatly reduced during assembly, especially compared to the process times of an assistance system for which the worker needs to switch process steps manually. Additionally, the acceptance of the assistance system increased with every extension level. The test person felt, for each extension level, that the assistance system was more of a support than a control device.

5   Conclusion

In this chapter, the development of an individual assembly system is presented. The worker assistance system includes an abstract representation for assembly flowcharts, a component for recording worker actions, and a representation unit for work instructions. The developed assistance system can support a worker so that he can successfully process and fulfil his assembly task.

The focus of this work is on three main tasks: the representation of the assembly schedule, the recording of the worker's activity, and the visualisation of the assistance information. A new concept for an individual assembly guidance system was shown. This assembly guidance system was implemented at an assembly station. The first step was to develop a system to automatically determine the process steps that are currently being carried out. For this purpose, several approaches, including 2D cameras, a pick-by-light system, and depth cameras, were used. Using a combined approach based mostly on camera images, an automatic determination of the process steps could be carried out. This assistance system with the automatic determination of process steps was the foundation of the development of the individual assembly guidance concept. For the individual assembly guidance system, five consecutive expansion stages of personalisation were defined. Personalisation is achieved by measuring a worker's process execution and adapting the instructions according to these measurements. With each step, the degree of freedom of the worker and thereby the personalisation of the approach were increased.

It could be shown that process times can be greatly reduced and the acceptance of assistance systems can be greatly improved by this approach. Additionally, this work shows that the personalisation of tasks through the usage of a digital twin is feasible.

References

Apt, W., Bovenschulte, M., Hartmann, E., & Wischmann, S. (2016). Foresight-Studie "Digitale Arbeitswelt" (Vol. Forschungsbericht 463). Bundesministerium für Arbeit und Soziales.

Arnold, D., & Furmans, K. (2009). *Materialfluss in logistiksystemen*. Berlin, Heidelberg: Springer. ISBN 978-3-642-01404-8, https://doi.org/10.1007/978-3-642-01405-5

Bannat, A. (2014). Ein Assistenzsystem zur digitalen Werker-Unterstützung in der industriellen Produktion. München.

Blumberg, V., & Kauffeld, S. (2020). Anwendungsszenarien und Technologiebewertung von digitalen Werkerassistenzsystemen in der Produktion–Ergebnisse einer Interview-Studie mit Experten aus der Wissenschaft, der Politik und der betrieblichen Praxis. *Gruppe Interaktion Organisation Zeitschrift für Angewandte Organisationspsychologie (GIO), 51*. https://doi.org/10.1007/s11612-020-00506-0

Bornewasser, M., Bläsing, D., & Hinrichsen, S. (2018). Informatorische Assistenzsysteme in der manuellen Montage: Ein nützliches Werkzeug zur Reduktion mentaler Beanspruchung? *Zeitschrift für Arbeitswissenschaft, 72*(4), 264–275. https://doi.org/10.1007/s41449-018-0123-x

Dangelmaier, W. (2009). *Theorie der Produktionsplanung und -steuerung: Im Sommer keine Kirschpralinen?* VDI-Buch: Springer.

DIN Deutsches Institut für Normung e V Normenausschuss Ergonomie. (2002). EN ISO 9241. Ergonomische Anforderungen für Bürotätigkeiten mit Bildschirmgerten.

Glonegger, M., Ottmann, W., Schadl, M., & Distel, D. (2013). Identification of production rhythms in synchronised assembly lines by recording and evaluating current processing times. In *WGP Congress 2013, Trans Tech Publications Ltd, Advanced Materials Research* (Vol 769, pp. 350–358). https://doi.org/10.4028/www.scientific.net/AMR.769.350

Graessler, I., & Poehler, A. (2018). Intelligent control of an assembly station by integration of a digital twin for employees into the decentralized control system. *Procedia Manufacturing, 24*, 185–189. https://doi.org/10.1016/j.promfg.2018.06.041

Hinrichsen, S., Bendzioch, S., Nikolenko, A., & Voss, E. (2018). Einsatzpotenziale von Montageassistenzsystemen. HOB, Holzbearbeitung (11).

Kagermann, H., Lukas, W. D., & Wahlster, W. (2011). Industrie 4.0: Mit dem internet der Dinge auf dem Weg zur 4. Industriellen revolution. *VDI Nachrichten, 13*(1), 2–3.

Klapper, J., Gelec, E., & Pokorni, B. (2020). Potenziale digitaler Assistenzsysteme: Aktueller und zukünftiger Einsatz digitaler Assistenzsysteme in produzierenden Unternehmen. Fraunhofer IAO.

Padoy, N., Mateus, D., Weinland, D., Berger, M. O., & Navab, N. (2009). Workflow monitoring based on 3D motion features. In *2009 IEEE 12th International Conference on Computer Vision Workshops, ICCV Workshops* (pp 585–592). https://doi.org/10.1109/ICCVW.2009.5457648

Schreiber, W. (2017). *Web-basierte Anwendungen Virtueller Techniken: Das ARVIDA-Projekt-Dienste-basierte Software-Architektur und Anwendungsszenarien für die Industrie*. Berlin, Heidelberg: Springer Vieweg.

Verband deutscher Ingenieure. (1990). *VDI 2860: Montage- und Handhabungstechnik; Handhabungsfunktionen, Handhabungseinrichtungen; Begriffe, Definitionen, Symbole*. Berlin: Beuth Verlag.

Zäh, M., & Reinhart, G. (2014). Assistenzsysteme in der Produktion. Werkstattstechnik Online, p. 516.

Integration of Human Factors
for Assembly Systems of the Future

Daniel Roesmann and Iris Gräßler

Abstract Assembly is the final step of production and has to adapt to changing requirements. Produced parts are assembled into a product of higher complexity with defined functions within a determined time. Workers are the central actors in future cyber-physical assembly systems. They are crucial to the success of the entire system. There is a variety of methods and models for planning specific aspects of assembly systems. Examples include workstation design, assembly layout, and task assignment. In these approaches, individual characteristics and human factors are insufficiently considered. Within this chapter, an approach for the integration of human factors into cyber-physical assembly systems is proposed. This approach is an extension for planning methods and models that is meant to optimise the performance and cost of the assembly system.

Keywords Assembly planning · Human factors · Line balancing · Worker assignment

1 Introduction

The manufacturing industry has been confronted with the production of customised products, increasing product complexity, and shorter development times. Companies have to react to market fluctuations quickly and also need to enable short-term order changes. Furthermore, companies are confronted with decreasing batch sizes but an increasing number of variants. In particular, the assembly process is challenged by a higher product variance (Argyrou et al., 2016). The task of assembly is to assemble a system of higher complexity with predefined functions from the produced individual parts within a defined period of time. The share of costs caused by assembly in production is up to 70%, and this confirms the enormous importance of assembly in the

D. Roesmann (✉) · I. Gräßler
Heinz Nixdorf Institute, Paderborn University, Fürstenallee 11, 33102 Paderborn, Germany
e-mail: daniel.roesmann@hni.uni-paderborn.de

I. Gräßler
e-mail: iris.graessler@hni.uni-paderborn.de

production industry (Lotter, 2012). Dombrowski et al. (2011) listed the following requirements for assembly systems: capacity utilisation, participation, synchronisation, costs, interfaces, networks, and cooperation, as well as ergonomic work design. Due to the flexible cognitive and motor skills of humans, most assemblies can only be operated in an economically sensible way with humans (Grosse, 2015; Stoessel et al., 2008). This is reflected in the high proportion of manual work, for example, in electrical and precision engineering (Kratzsch, 2000). In order to adapt to market-driven changes in assembly, a flexible and adaptable workforce is required. Many studies (Adolph et al., 2014; Lanza et al., 2019; Shalin et al., 1996; Stoessel et al., 2008) have recommended a human-centred approach to assembly planning as a solution. Multi-skilled workers can adapt to new and changing tasks after a learning process (Brettel et al., 2014).

In parallel, the development of information and communication technologies has an influence on production engineering research (Hozdic, 2015). Digitalisation offers new applications such as the Internet of Things, digital twin, big data analytics, predictive maintenance, and additive manufacturing. The design of production systems as cyber-physical systems enables the creation of cyber-physical production systems and smart factories (Neumann et al., 2021). These link the physical components with digital and virtual models of the production system. The main objective is to create and improve transparency in the production system and to create the possibility of real-time production control. However, most planning models that exist to support decision-making in assembly systems neglect the specific characteristics of human workers (Sgarbossa et al., 2020). Neumann and Dul (2010) claimed 'that careful application of Human Factors principles in the design of operations can improve productivity, quality, technology implementation, and have in-tangible benefits for operations while also securing well-being and working conditions for employees.' In the following, a framework for how the digital twin of a human being can improve the assembly process by scheduling assembly processes in cyber-physical assembly systems is shown. The digital twin of a human being is a digital representation that contains the selected characteristics and behaviours of a person. Therefore, existing validated models for the consideration of human factors are integrated into assembly scheduling.

2 Working in Cyber-Physical Assembly Systems

To compete in the global market and to be successful in the long term, it is necessary to implement digital information and communication technologies in production systems, processes, and technologies. Cyber-physical production systems consist of autonomous and cooperative elements and subsystems that interact depending on the situation across all levels of production, from processes to machines to production and logistics networks (Chatti et al., 2019). According to Monostori et al. (2016), cyber-physical production systems have three main characteristics:

- *Intelligence* (smartness), i.e., the elements are able to acquire and process information from their surroundings, determine their own state, and act autonomously;
- *Connectedness*, i.e., the ability to set up and use connections to other elements of the system—including human beings—for cooperation and collaboration, and to use connections to the knowledge and services available on the internet;
- *Responsiveness* towards internal and external changes.

Assembly processes are upgraded in the same way in order to adapt to these requirements. The resulting socio-technical systems are characterised by a high degree of networking of the physical, virtual, and social worlds and by the intelligent use of information and communication systems (Geisberger & Broy, 2012). The use of cyber-physical production systems technologies in assembly environments enables the creation of so-called cyber-physical assembly systems (Galaske, 2019). The system model includes products, processes, resources, and workers, as well as assembly stations, individual assembly machines, and equipment (Gräßler & Pöhler, 2017, 2018; Gräßler et al., 2017, 2021b; Hammerstingl & Reinhart, 2018; Müller et al., 2019; Strang, 2016):

- The *product model* consists of the product master data and product structure, as well as documents and document structures. The parts lists describe the allocation of product components (materials, individual parts, assemblies, products).
- The *process model* represents the logical and/or chronological procedure of the assembly process. Therefore, among other things, assembly precedence graphs are used. These graphs list the assembly processes with corresponding predecessor/successor relationships.
- The *assembly resources model* includes all the resources of the assembly system that are used to assemble the product. These resources are, for example, workstations, assembly tools, and machines.
- The *worker model* contains information about the workers in the assembly system. This is, on the one hand, general information such as the worker's identification number and availability. On the other hand, the representation involves information about specific human factors. According to Grosse et al. (2015), human factors can be subdivided into perceptual, mental, physical, and psychosocial aspects, as well as work environment aspects.

Within a cyber-physical assembly system, the static data of the assembly system are connected to real-time data. The system elements are extended by suitable sensors, communication interfaces, and actuators, which enable the linking of the digital and physical worlds. At every element, relevant data about the product, process, resources, and workers are recorded in a machine-readable way. This includes current states, such as the number of assemblies, the abrasion of resources, or worker experience. These data are exchanged via a communication interface. The information processing system receives the data input. The connection of the cyber-physical assembly system with the Internet of Things and Services enables, for example, the integration of artificial intelligence services for predictive analytics. Optimisation activities for the system can be determined in real time and passed via actors (Strang & Anderl, 2014).

Another cornerstone of the planning of the assembly systems of the future is understanding the influence of human factors and integrating them into socio-technical system development (Brauner & Ziefle, 2015; Graessler & Poehler, 2019). System designers have to consider the cognitive, emotional, and motivational needs and skills, as well as physical and anthropometric characteristics, of the workers (Wilkowska & Ziefle, 2013). This includes the effects of an ageing workforce and questions about the extent to which workers with different qualifications will learn and adapt to the increasing complexity of systems. To this end, workers and their knowledge must be taken into account as an integral part of the development of new systems at an early stage. The interdisciplinary view of human factors deals with the relationship between people and technology from a systematic perspective. The study of human factors is based on the disciplines of engineering, management science, mathematics, computer science, education, ergonomics, law and occupational science, and psychology (Badke-Schaub et al., 2012). The aim is to design and evaluate tasks, work, products, environments, and systems to meet the needs, skills, and limitations of the workers. In industrial assembly, decisions have a significant influence on the worker and the productivity of the entire socio-technical system. For example, worker assignment within scheduling has the following three exemplary aspects:

- *Physical aspects:* If the workers are always assigned to the same station, the workers are physically strained. If the station requires heavy objects to be lifted, the worker's back and joints will be strained in the long run. This leads to a drop in performance but above all to the physical impairment of the respective worker (Hochdörffer et al., 2018).
- *Psychological aspects:* Frequent changes between workplaces and tasks lead to a varied working day for the workers. This can keep workers focused and motivated, which in turn increases the productivity and effectiveness of the assembly system (Arnold, 1997).
- *Learning aspects:* Within assembly, work-integrated learning is of particular importance. The development of skills takes place as part of the daily work process and can be influenced by assigning workers to different assembly tasks. Continuous work-integrated learning is necessary for demand-oriented and efficient qualification (Gräßler et al., 2021a).

3   Assembly Planning

Industrial assembly is essentially shaped by current market influences. In the literature, there exist many different methods that support assembly planning. The basic procedures for manual assembly are the 6-step method of REFA (1990), the VDI Guidelines (Verein Deutscher Ingenieure, 2019, 2020), and the methods of Bullinger and Ammer (1986), Lotter (1992), Feldmann (1997), Grunwald (2002), and Patron (2005). The planning procedure of REFA (1990) serves as the basis for the others. This procedure is briefly described as follows. Based on an analysis of the initial

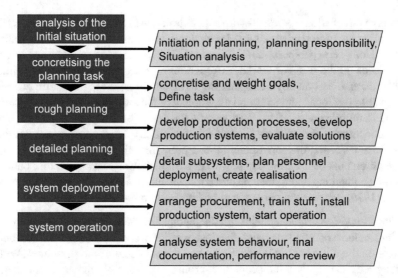

Fig. 1 The 6-step method of REFA (1990)

situation, the objectives are specified and the individual tasks are defined. Then, the rough planning takes place; this involves designing the relevant production processes. This is the basis for planning individual workstations. Within the framework of personnel deployment planning, worker requirements and qualification needs are derived. Finally, the system is introduced and put into operation (REFA, 1990) (see Fig. 1).

A central step in the planning of assembly lines is scheduling and balancing. An assembly line consists of workstations that can be arranged along a conveyor belt. The workpieces are moved from station to station. At each station, certain assembly operations are performed with respect to the cycle time for each individual product (Kriengkorakot & Pianthong, 2007). Therefore, scheduling is the assignment of limited capacities to a set of tasks in terms of time and quantity. The timing of orders and capacities is determined on the basis of product-specific work plans. Furthermore, the chronological sequence of assembly operations is carried out in compliance with technical specifications, i.e., chronologically, in parallel, or in sequence. Classical optimisation procedures cover the following objectives: the minimisation of order throughput times, minimisation of process-related machine downtimes, uniform capacity utilisation, and minimisation of delay times. An important lean production technique within the scheduling of assembly lines is line balancing. Line balancing is a technique for minimising the imbalance between workloads in order to achieve the required run rate. A balanced line increases the efficiency and productivity of an assembly line (Ege et al., 2009). The classical assembly line balancing problem considers the assignment of tasks to workstations. The goal is to minimise the total assembly cost while satisfying the requirements and restrictions of the assembly station (Becker & Scholl, 2006). To solve these problems, so-called advanced planning and scheduling systems with a module called Production Planning/Detailed Scheduling are used (Betge, 2006). The assignment of tasks can be done from short-term and long-term perspectives (Gräßler et al., 2021a). The determination of suitable workplaces on the basis of worker information represents a personnel deployment problem. Personnel deployment planning in assembly involves the future quantitative, qualitative, local, and temporal assignment of the available personnel capacities in the assembly process, taking into account the operational objectives and the legitimate concerns of the workers (Schuh & Stich, 2012). Thus, workers are assigned to perform tasks and the corresponding orders by matching the workers' skills with the requirements of the task. Thus, worker assignment represents the link between production planning and human resource management. According to Eversheim and Schuh (1999), worker assignment can be divided into two problems:

1. The *adaptation problem* deals with the mutual adaptation of personnel and work. This includes, on the one hand, the design of the work tasks to suit the personnel and, on the other hand, the promotion and development of the personnel. With a long-term orientation, the adaptation problem in assembly planning is meant to encourage the development of the worker through work-integrated learning.
2. The *assignment problem* is divided into quantitative assignment and qualitative assignment. This involves the appropriate quantitative assignment of workers to work tasks in order to create shift schedules. These schedules contain information on the production quantity in each shift and how many workers are assigned to each shift. Qualitative approaches extend this procedure by matching each worker's competence profile with the requirement profile of each work task. By checking the workers' suitability, the product quality and occupational safety can be ensured.

4 Modelling of Worker for Assembly Planning

Models for the consideration of the worker already exist in the literature. Researchers such as Gräßler et al. (2021), Abubakar and Wang (2019b), and Katiraee et al. (2021) presented overviews of research on the consideration of human differences in assembly and production. This research shows that the human factors *learning, forgetting,* and *ageing* have a decisive influence on the performance of an assembly system. In assembly, a learning process causes a worker to need less time to perform the assembly task. In contrast, forgetting and ageing result in the worker needing a longer amount of time (see Fig. 2). The fluctuations in the processing time are not purely random; they can be explained to a certain extent. For selected characteristics, existing modelling techniques are presented in the following subsections. These models can be integrated into the overall system model of an assembly line and can be used in discrete event simulations.

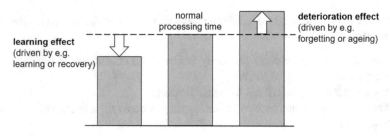

Fig. 2 Effects of learning and deterioration on processing times according to Ostermeier (2020)

4.1 Learning

Due to repetitions of operations, an informal learning process takes place. This leads to increasing experience and decreasing assembly times (Anzanello & Fogliatto, 2006). This concept is based on the observations of Wright (1936) of the production of aircraft in the 1930s. He noticed that the processing time increased by only 80% when the output was doubled. This reduction in the processing time followed a constant rate and led to the '80% learning curve' in the aeronautical industry. Learning can be considered in assembly planning by integrating the concept of learning curves, which makes it possible to model the performance improvement of human workers over time. Learning curves assume that the performance improves as a task is repetitively performed (Grosse, 2015). Workers tend to require less time to perform tasks due to their familiarity with the operation and tools. The learning theories in production led to a statistically proven regularity of learning. Learning curves are often used in decision support systems, for example, to determine the production rate (Jaber, 2006), to select suppliers in commissioning systems (Stinson et al., 2016), or to determine an appropriate task assignment (Gräßler et al., 2021a). A variety of different learning curve models exist. They are divided into univariate and multivariate learning curves. The best-known univariate learning curves include the logarithmic-linear, exponential, and hyperbolic models. For a detailed overview of existing learning curves, see Anzanello and Fogliatto (2011).

4.2 Learning and Forgetting

The opposite of learning is forgetting. If a certain action has not been periodically carried out, a worker may, for example, forget the process sequence of the assembly operation. This phenomenon has been observed and researched. Bailey (1989), Globerson et al. (1989), and Nembhard and Uzumeri (2000) conducted experimental studies in laboratory settings to investigate the connection between learning and forgetting. They have determined that forgetting occurs in one of the following situations (Jaber, 2006):

1. When there is not enough similarity between the conditions of encoding and retention of the material learned,
2. When old learning interferes with new learning,
3. When there is an interruption in the learning process for a period of time (production break).

In contrast to the amount of literature on learning curves, there is a lack of literature on forgetting curves. This lack of research has been attributed to the practical difficulties involved in obtaining data concerning forgetting as a function of time (Jaber, 2006). Regarding the mathematical description of this process, there are models in the literature that consider experience, learning, and forgetting called learning and forgetting models. Well-known and common models include the learn-forget curve model from Jaber and Bonney (1997), the recency model from Nembhard and Uzumeri (2000), the power integration diffusion model from Sikström and Jaber (2002), and the modified learn-forget curve model from Jaber and Kher (2004). These models can be incorporated into assembly planning. With the help of skill development predictions, scheduling forecasts can be improved.

4.3 Age

Today, companies are facing ageing workforces in assembly systems (Katiraee et al., 2019). In addition to the increasing experience of older workers, there are age-related changes in their capabilities. Ageing leads to human physiological decline, which affects a person's individual performance. This phenomenon was observed in many studies. For example, Shephard (2000) reported that age affects the occupational performance of older people due to a decline in aerobic capacity. Goebel and Zwick (2009) found that the average muscle strength of a person that is 20–60 years old decreases by about 10% per decade. It then decreases by 15% between 60 and 80 years old and by 30% after 80 years old. Based on 16 empirical studies, Abubakar and Wang (2019a) developed a mathematical model of capability as a function of age within an assembly context. To do this, they referred to studies that consider the aerobic capacity, spatial capability, flexibility, and overall physiological capability of workers. The result is that the highest level of the worker's functional ability is reached at the age of 38, and thereafter, a constant reduction in this ability takes place.

5 Human-Centred Modelling to Support Decisions in Assembly Systems

Within assembly planning and control, decision support systems, from optimised, knowledge-based models to simulation-based models, focus on the design and management of assembly systems. Through efforts such as those of the *NRW Research College: Design of Flexible Work Environments—Human Centric Use of Cyber-Physical Systems in Industry 4.0* group and the working group *Human Factors and Ergonomics in Industrial and Logistic System Design and Management* of the IFAC Technical Committee, human-centred approaches are gaining importance in the modelling of socio-technical production systems in industrial engineering. The development and implementation of cyber-physical assembly systems enables the digital twins of humans to be used in assembly planning and control. For this purpose, based on the generic approach of Sgarbossa et al. (2020), a specific framework for human-centred decision support systems in assembly systems that uses the human digital twin was developed within this chapter. The framework is shown in Fig. 3. It consists of human-centred assembly planning, control, human-centred modelling, operations research, human-centred assembly management, and data science.

Human-centred assembly planning: It is crucial that the workers are taken into account in the design of an assembly system. For this purpose, existing assembly line balancing problems from the literature must be expanded, for example, to enable

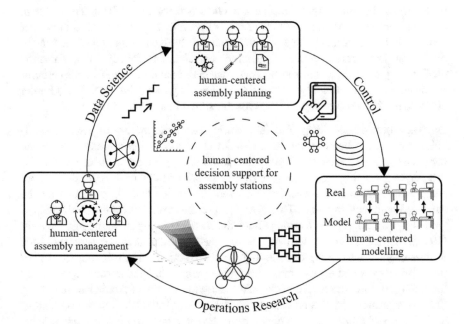

Fig. 3 Human-centred decision support for assembly systems (adapted from Sgarbossa et al., 2020)

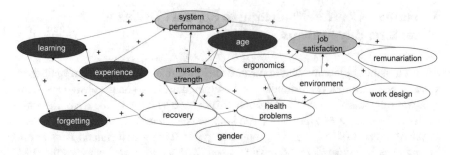

Fig. 4 Effects of human factors on assembly system performance (Gräßler et al., 2021)

individually designed workplaces. Boysen et al. (2007) provided a good overview of existing types of approaches for solving different assembly line balancing problems. For a human-centred design, the problem is expanded by integrating human factors into the decision-making process. In a meta-analysis, Gräßler et al. (2021) found that the human factors experience, age, learning, and forgetting have the highest influence on the system performance (see Fig. 4). Therefore, it is crucial for assembly balancing to take these factors into account.

Control: Within the execution of the work, the use of information and communication technologies in cyber-physical assembly systems enables the capture and management of large amounts of data to improve the understanding of the system under investigation. For this purpose, digital assistance systems are used. On the one hand, the assistance system visualises the necessary assembly steps for the worker at their station. On the other hand, the assistance system is used as an input device. Among other data, the workers' process times are recorded. The collected data form the worker's digital shadow. The digital assistance system at the assembly stations communicates with the assembly planning and scheduling tool using OPC UA (Gräßler et al., 2016, 2020a) (see Fig. 5). The data are stored in a common database.

Human-centred modelling: Within assembly planning, the use of discrete simulation models has been established. Gräßler et al. (2021a) considered the human factors experience, ageing, learning, and forgetting in a mathematical model. For this purpose, existing models from occupational science studies were used. The core component is the *learning and forgetting model*. The MLFCM in particular can be enhanced with real-time data. Therefore, it represents the basis for the mathematical modelling. To describe the learning effect, the classic learning curve model of Wright (1936) is used. This is extended using the findings of Pleteau models (as presented, for example, by Ullrich (1995) for assembly planning) concerning a minimum assembly time. This means that the assembly time runs against a defined minimum execution time. Furthermore, the model is extended using the results of Abubakar and Wang (2019a) regarding age-related changes in skills. In order to take into account similar tasks in addition to the experience from the same task, the factor 'concordance

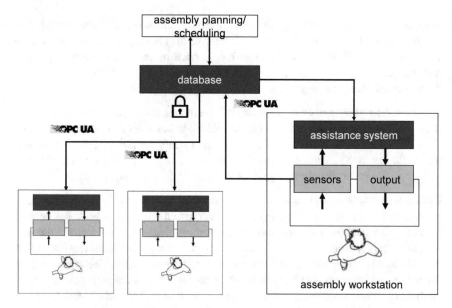

Fig. 5 Data communication in cyber-physical assembly systems

of assembly processes' is considered in the modelling. Accordingly, the following learning curve is used:

$$t(n, i) = t_{min} + ((t(1, 1) - t_{min}) * (n + \sum_{x=1}^{X} Acc_{s_x s_y} * u_{is_x}))^{(} - b) * (1 - Vr * (k - 38)),$$

(1)

where $t(n, i)$ is the time to assemble the nth unit in the ith cycle, b is the learning rate, $Acc_{s_x s_y}$ is the similarity between skill s_i from one assembly station and skill s_j from another assembly station, u_{is_x} represents the recalled repetitions at the assembly stations after an interruption in the ith cycle, Vr is the loss rate, and k is the age. The forgetting curve is modelled in a similar way:

$$\hat{t}(n, i) = \hat{t}(1, i) * n^f,$$

(2)

where $\hat{t}(n, i)$ is the time to assemble the nth unit in the forgetting curve. n represents the number of units that would have been completed without the interruption. f is the forgetting rate.

Operations research: Operations research methods are integrated into the model for systematic analysis and problem solving. The objective of operations research is to solve the optimisation problem as accurately as possible, taking into account the relevant constraints (Klein & Scholl, 2012; Neumann & Morlock, 2004). For this purpose, according to Gräßler et al. (2021b), different subfields of operations research are useful: linear programming, graph theory, integer and combinatorial optimisation, dynamic optimisation, non-linear optimisation, queueing theory, and simulation (Domschke et al., 2015). Gräßler et al. (2021a) suggested the use of the Hungarian method (Kuhn, 1955) for long-term skill-oriented worker assignment and the use of the bottleneck allocation model of Pundir et al. (2015) for short-term time-oriented worker assignment.

Human-centred assembly management: The results from operations research are used to support and improve the decision-making behaviour or to forecast and react to it. The acceptance and success of decision support depend on the data quality, interpretation mechanisms, transparency, accuracy, controllability, and traceability. Therefore, tools for visual management, such as Andon boards, are used.

Data science/data analytics: Data science/data analytics methods are used to analyse the data generated within a cyber-physical assembly system. With the help of data analytics methods from the areas of descriptive, diagnostic, predictive, and prescriptive analytics, analyses can be carried out. In order to obtain conclusions about the individual development of skills, a regression analysis is used. This approximates the individual learning curve of the workers on the basis of discrete data. After an analysis of existing methods, the least squares method was chosen. The determined learning outcomes are checked in terms of correctness, for example range of value, previous results, and incorporated into the forecast. In this way, the digital twin of a human is continuously specified (Gräßler et al., 2020b).

6 Conclusion

Although there is a wide range of technical possibilities for automating assembly processes, assembly in many production sectors is still predominantly carried out manually, with a high proportion of human workers. The advantage of human workers in assembly is their flexibility in responding to changes in the work environment, new products, and new technologies. This is possible due to the combination of human cognitive and motor skills; human workers cannot be economically replaced by machines. Assembly systems must therefore be considered, planned, and controlled from a socio-technical perspective. The specific requirements of the workers must be taken into account in the development of the system. This especially applies to new assembly systems supported by digital tools. The permeation of production by information and communication technology enables new design possibilities, such as the use of the Internet of Things and Services or big data. Today, the planning and control of such cyber-physical assembly systems is often model-based. Within these models, the workers, along with their individual characteristics, can be linked in human-centred assembly planning. In this way, the relevant characteristics can be considered. Modelling can be carried out on the basis of occupational science and sociological findings. Building on this, this chapter presented a framework for the human-centred decision support of assembly stations. Within a generic approach, recommendations for the technical implementation of specific issues are given. The entire framework consists of six elements: human-centred assembly planning, control, human-centred modelling, operations research, human-centred assembly management, and data science.

Acknowledgements Daniel Roesmann and Iris Gräßler are members of the research programme 'Design of Flexible Work Environments—Human-Centric Use of Cyber-Physical Systems in Industry 4.0', which is supported by the North Rhine-Westphalian funding scheme 'Forschungskolleg'.

References

Abubakar, M. I., & Wang, Q. (2019a) Integrating human factor decision components into a DES model. In *2019 IEEE 6th International Conference on Industrial Engineering and Applications (ICIEA)* (pp. 562–566). IEEE. https://doi.org/10.1109/IEA.2019.8714774

Abubakar, M. I., & Wang, Q. (2019b). Key human factors and their effects on human centered assembly performance. *International Journal of Industrial Ergonomics, 69*, 48–57. https://doi.org/10.1016/j.ergon.2018.09.009

Adolph, S., Tisch, M., & Metternich, J. (2014). Challenges and approaches to competency development for future production. *Journal of International Scientific Publications Educational Alternatives, 12*(1), 1001–1010.

Anzanello, M. J., & Fogliatto, F. S. (2006). Learning curve modelling of work assignment in mass customized assembly lines. *International Journal of Production Research, 45*(13), 2919–2938. https://doi.org/10.1080/00207540600725010

Anzanello, M. J., & Fogliatto, F. S. (2011). Learning curve models and applications: Literature review and research directions. *International Journal of Industrial Ergonomics, 41*(5), 573–583. https://doi.org/10.1016/j.ergon.2011.05.001

Argyrou, A., Giannoulis, C., Papakostas, N., & Chryssolouris, G. (2016). A uniform data model for representing symbiotic assembly stations. *Procedia CIRP, 44*, 85–90. https://doi.org/10.1016/j.procir.2016.02.087

Arnold, W. (Ed.). (1997). *Lexikon der Psychologie*. Augsburg: Bechtermünz.

Badke-Schaub, P., Hofinger, G., & Lauche, K. (2012). *Human factors*. Dordrecht: Springer.

Bailey, C. D. (1989). Forgetting and the learning curve: A laboratory study. *Management Science, 35*(3), 340–352. https://doi.org/10.1287/mnsc.35.3.340

Becker, C., & Scholl, A. (2006). A survey on problems and methods in generalized assembly line balancing. *European Journal of Operational Research, 168*(3), 694–715. https://doi.org/10.1016/j.ejor.2004.07.023

Betge, D. (Ed.). (2006). *Koordination in advanced planning and scheduling-systemen* (1st edn.) Produktion und Logistik, DUV Deutscher Universitäts-Verlag, s.l. https://doi.org/10.1007/978-3-8350-9041-5

Boysen, N., Fliedner, M., & Scholl, A. (2007). A classification of assembly line balancing problems. *European Journal of Operational Research, 183*(2), 674–693. https://doi.org/10.1016/j.ejor.2006.10.010

Brauner, P., & Ziefle, M. (2015). Human factors in productions systems. Lecture Notes in Production EngineeringIn C. Brecher (Ed.), *Advances in production technology* (pp. 187–199). Cham: Springer Open.

Brettel, M., Friederichsen, N., Keller, M., & Rosenberg, N. (2014). How virtualization, decentralization and network building change the manufacturing landscape: An industry 4.0 perspective. *International Journal of Science, Engineering and Technology, 8*, 37–44. https://doi.org/10.5281/zenodo.1336426

Bullinger, H. J., & Ammer, D. (Eds.). (1986). *Systematische Montageplanung: Handbuch für die Praxis*. München: Hanser. ISBN 978-3-446-14606-8.

Chatti, S., Laperrière, L., & Reinhart, G. (2019). *CIRP encyclopedia of production engineering* (2nd ed.). Berlin: Springer. https://doi.org/10.1007/978-3-642-20617

Dombrowski, U., Riechel, C., & Schulze, S. (2011). Enforcing employees participation in the factory planning process. In *2011 International Symposium on Assembly and Manufacturing* (pp. 1–6). Piscataway, NJ: IEEE. https://doi.org/10.1109/ISAM.2011.5942337

Domschke, W., Drexl, A., Klein, R., & Scholl, A. (2015). *Einführung in operations research* (9th edn.). Berlin and Heidelberg: Springer Gabler. https://doi.org/10.1007/978-3-662-48216-2

Ege, Y., Azizoglu, M., & Ozdemirel, N. E. (2009). Assembly line balancing with station paralleling. *Computers & Industrial Engineering, 57*(4), 1218–1225. https://doi.org/10.1016/j.cie.2009.05.014

Eversheim, W., & Schuh, G. (1999). *Produktion und Management 3: Gestaltung von Produktionssystemen*. Berlin and Heidelberg: Hütte, Springer. https://doi.org/10.1007/978-3-642-58399-5

Feldmann, C. (1997). *Eine Methode für die integrierte rechnergestützte Montageplanung, Forschungsberichte iwb, Berichte aus dem Institut für Werkzeugmaschinen und Betriebswissenschaften der Technischen Universität München* (Vol. 104). Berlin and Heidelberg: Springer. https://doi.org/10.1007/978-3-662-06845-8

Galaske, N. R. (2019). *Modellierung von Zusammenbaubedingungen zur Reihenfolgebildung im cyber-physischen Montagesystem, Forschungsberichte aus dem Fachgebiet Datenverarbeitung in der Konstruktion* (1st ed., Vol. 63). Herzogenrath: Shaker. ISBN 978-3-8440-6523-7

Geisberger, E., & Broy, M. (2012). *agendaCPS: Integrierte Forschungsagenda Cyber-Physical Systems, acatech STUDIE, März 2012* (Vol. 1). Berlin and Heidelberg: Springer. https://doi.org/10.1007/978-3-642-29099-2

Globerson, S., Levin, N., & Shtub A. (1989). The impact of breaks on forgetting when performing a repetitive task. *IIE Transactions, 21*(4), 376–381. https://doi.org/10.1080/07408178908966244

Goebel, C., & Zwick, T. (2009). *Age and productivity—Evidence from linked employer employee data*. Labor: Personnel Economics. https://ftp.zew.de/pub/zew-docs/dp/dp09020.pdf

Graessler, I., & Poehler, A. (2019). Human-centric design of cyber-physical production systems. *Procedia CIRP, 84*, 251–256. https://doi.org/10.1016/j.procir.2019.04.199

Gräßler, I., & Pöhler, A. (2017). Implementation of an adapted holonic production architecture. *Procedia CIRP, 63*, 138–143. https://doi.org/10.1016/j.procir.2017.03.176

Gräßler, I., Pöhler, A., & Pottebaum, J. (2016). Creation of a learning factory for cyber physical production systems. *Procedia CIRP, 54*, 107–112. https://doi.org/10.1016/j.procir.2016.05.063

Gräßler, I., Roesmann, D., & Pottebaum, J. (2020a). Entwicklung eines Prüfstands für die Bewertung von kompetenzbildenden Assistenzsystemen in cyber-physischen Produktionssystemen. In *Digitale Arbeit, digitaler Wandel, digitaler Mensch?* (p. B.6.4). Dortmund: GfA-Press. ISBN 978-3-936804-27.

Gräßler, I., Roesmann, D., & Pottebaum, J. (2020b). Traceable learning effects by use of digital adaptive assistance in production. In *Proceedings of the 10th Conference on Learning Factories* (pp. 479–484). Elsevier. https://doi.org/10.1016/j.promfg.2020.04.058

Gräßler, I., Roesmann, D., Cappello, C., & Steffen, E. (2021). Skill-based worker assignment in a manual assembly line. *Procedia CIRP, 100*, 433–438. https://doi.org/10.1016/j.procir.2021.05.100

Gräßler, I., Wiechel, D., & Roesmann, D. (2021). Integrating human factors in the model based development of cyber-physical production systems. *Procedia CIRP, 100*, 518–523. https://doi.org/10.1016/j.procir.2021.05.113

Gräßer, I., & Pöhler, A. (2018). Intelligent devices in a decentralized production system concept. *Procedia CIRP, 67*, 116–121. https://doi.org/10.1016/j.procir.2017.12.186, www.sciencedirect.com/science/article/pii/S2212827117311289

Gräßler, I., Pöhler, A., & Hentze, J. (2017). Decoupling of product and production development in flexible production environments. *Procedia CIRP, 60*, 548–553. https://doi.org/10.1016/j.procir.2017.01.040

Gräßler, I., Roesmann, D., & Pottebaum, J. (2021). Model based integration of human characteristics in production systems: A literature survey. *Procedia CIRP, 99*, 57–62. https://doi.org/10.1016/j. procir.2021.03.010, www.sciencedirect.com/science/article/pii/S2212827121002663

Grosse, E. H. (2015). Human factors in order picking systems: A framework for integrating human factors in order picking planning models with an in-depth analysis of learning effects. Dissertation, Technische Universität Darmstadt, Darmstadt. ISBN 978-3-944325-05-7.

Grosse, E. H., Glock, C. H., Jaber, M. Y., & Neumann, W. P. (2015). Incorporating human factors in order picking planning models: Framework and research opportunities. *International Journal of Production Research, 53*(3), 695–717. https://doi.org/10.1080/00207543.2014.919424

Grunwald, S. (2002). Methode zur Anwendung der flexiblen integrierten Produktentwicklung und Montageplanung: Zugl.: München, Technical University, Dissertation (2001). *Forschungsberichte/IWB* (Vol. 159). München: Utz. ISBN 978-3-8316-0095-3.

Hammerstingl, V., & Reinhart, G. (2018). Skills in assembly.

Hochdörffer, J., Hedler, M., & Lanza, G. (2018). Staff scheduling in job rotation environments considering ergonomic aspects and preservation of qualifications. *Journal of Manufacturing Systems, 46*, 103–114. https://doi.org/10.1016/j.jmsy.2017.11.005

Hozdic, E. (2015). Smart factory for industry 4.0: A review. *International Journal of Modern Manufacturing Technologies, 7*(1), 28–35. ISSN 2067–3604.

Jaber, M. (2006). Learning and forgetting models and their applications. In A. B. Badiru (Ed.), *Handbook of industrial and systems engineering, Industrial innovation* (Vol. 20052471, pp. 30–1–30–27). Boca Raton: CRC Taylor & Francis. https://doi.org/10.1201/9781420038347.ch30

Jaber, M. Y., & Bonney, M. (1997). A comparative study of learning curves with forgetting. *Applied Mathematical Modelling, 21*(8), 523–531. https://doi.org/10.1016/S0307-904X(97)00055-3

Jaber, M. Y., & Kher, H. V. (2004). Variant versus invariant time to total forgetting: The learn-forget curve model revisited. *Computers & Industrial Engineering, 46*(4), 697–705. https://doi.org/10.1016/j.cie.2004.05.006

Katiraee, N., Battini, D., Battaia, O., & Calzavara, M. (2019). Human diversity factors in production system modelling and design: State of the art and future researches. *IFAC-PapersOnLine, 52*(13), 2544–2549. https://doi.org/10.1016/j.ifacol.2019.11.589

Katiraee, N., Calzavara, M., Finco, S., Battini, D., & Battaïa, O. (2021). Consideration of workers' differences in production systems modelling and design: State of the art and directions for future research. *International Journal of Production Research, 59*(11), 3237–3268. https://doi.org/10.1080/00207543.2021.1884766

Klein, R., & Scholl, A. (2012). *Planung und Entscheidung: Konzepte, Modelle und Methoden einer modernen betriebswirtschaftlichen Entscheidungsanalyse.* Vahlens Handbücher der Wirtschafts- und Sozialwissenschaften, München. ISBN 978-3-8006-3884-0

Kratzsch, S. (2000). Prozess- und Arbeitsorganisation in Fließmontagesystemen Braunschweig, Technical University, Dissertation (2000). *Schriftenreihe des IWF.* Essen: Vulkan-Verl. ISBN 978-3-8027-8654-9.

Kriengkorakot, N., & Pianthong, N. (2007). The assembly line balancing problem: Review articles. *Engineering and Applied Science Research, 34*(2), 133–140.

Kuhn, H. W. (1955). The Hungarian method for the assignment problem. *Naval Research Logistics Quarterly, 2*(1–2), 83–97. https://doi.org/10.1002/nav.3800020109

Lanza, G., Ferdows, K., Kara, S., Mourtzis, D., Schuh, G., Váncza, J., Wang, L., & Wiendahl, H. P. (2019). Global production networks: Design and operation. *CIRP Annals, 68*(2), 823–841. https://doi.org/10.1016/j.cirp.2019.05.008

Lotter, B. (1992). *Wirtschaftliche Montage: Ein Handbuch für Elektrogerätebau und Feinwerktechnik* (2nd ed.). Düsseldorf: VDI-Verl. ISBN 978-3184007096.

Lotter, B. (2012). Einführung. In B. Lotter & H. P. Wiendahl (Eds.), *Montage in der industriellen Produktion* (pp. 1–8). Berlin, Heidelberg: VDI-Buch, Springer. https://doi.org/10.1007/3-540-36669-5

Monostori, L., Kádár, B., Bauernhansl, T., Kondoh, S., Kumara, S., Reinhart, G., Sauer, O., Schuh, G., Sihn, W., & Ueda, K. (2016). Cyber-physical systems in manufacturing. *CIRP Annals, 65*(2), 621–641. https://doi.org/10.1016/j.cirp.2016.06.005

Müller, R., Hörauf, L., Speicher, C., & Obele, J. (2019). Communication and knowledge management platform for concurrent product and assembly system development. *Procedia Manufacturing, 28*, 107–113. https://doi.org/10.1016/j.promfg.2018.12.018

Nembhard, D. A., & Uzumeri, M. V. (2000). Experiential learning and forgetting for manual and cognitive tasks. *International Journal of Industrial Ergonomics, 25*(4), 315–326. https://doi.org/10.1016/S0169-8141(99)00021-9

Neumann, K., & Morlock, M. (2004). *Operations research* (2nd ed.). München: Hanser.

Neumann, P. W., & Dul, J. (2010). Human factors: Spanning the gap between OM and HRM. *International Journal of Operations & Production Management, 30*(9), 923–950. https://doi.org/10.1108/01443571011075056

Neumann, W. P., Winkelhaus, S., Grosse, E. H., & Glock, C. H. (2021). Industry 4.0 and the human factor—A systems framework and analysis methodology for successful development. *International Journal of Production Economics, 233*, 107992. https://doi.org/10.1016/j.ijpe.2020.107992

Ostermeier, F. F. (2020). The impact of human consideration, schedule types and product mix on scheduling objectives for unpaced mixed-model assembly lines. *International Journal of Production Research, 58*(14), 4386–4405. https://doi.org/10.1080/00207543.2019.1652780

Patron, C. (2005) Konzept für den Einsatz von Augmented Reality in der Montageplanung München, Technical University, Dissertation (2004). *Forschungsberichte iwb / Institut für Werkzeugmaschinen und Betriebswissenschaften der Technischen Universität München* (Vol. 190). München: Utz. ISBN 978-3-8316-0474-6.

Pundir, P. S., Porwal, S. K., & Singh, B. P. (2015). A new algorithm for solving linear bottleneck assignment problem. *Journal of Institute of Science and Technology, 20*(2), 101–102. https://doi.org/10.3126/jist.v20i2.13961

REFA. (1990). *Planung und Gestaltung komplexer Produktionssysteme, Methodenlehre der Betriebsorganisation, vol / REFA, Verband für Arbeitsstudien und Betriebsorganisation* (2nd ed.). München: Hanser. ISBN 978-3446159679.

Schuh, G., & Stich, V. (Eds.). (2012). *Produktionsplanung und -steuerung* (4th ed.). Berlin and Heidelberg: VDI-Buch, Springer Vieweg.

Sgarbossa, F., Grosse, E. H., Neumann, W. P., Battini, D., & Glock, C. H. (2020). Human factors in production and logistics systems of the future. *Annual Reviews in Control, 49*, 295–305. https://doi.org/10.1016/j.arcontrol.2020.04.007

Shalin, V. L., Prabhu, G. V., & Helander, M. G. (1996). A cognitive perspective on manual assembly. *Ergonomics, 39*(1), 108–127. https://doi.org/10.1080/00140139608964438

Shephard, R. J. (2000). Aging and productivity: Some physiological issues. *International Journal of Industrial Ergonomics, 25*(5), 535–545. https://doi.org/10.1016/S0169-8141(99)00036-0

Sikström, S., & Jaber, M. Y. (2002). The power integration diffusion model for production breaks. *Journal of Experimental Psychology: Applied, 8*(2), 118–126. https://doi.org/10.1037//1076-898X.8.2.118

Stinson, M. R., Müller, F. H., Korte, D., & Wehking, K. H. (2016). Lernkurven in manuellen person-zur-ware-kommissioniersystemen (leikom): Abschlussbericht. https://www.bvl.de/files/1951/2125/2131/2133/Abschlussbericht_LeiKom.pdf

Stoessel, C., Wiesbeck, M., Stork, S., Zaeh, M. F., & Schuboe A. (2008). Towards optimal worker assistance: Investigating cognitive processes in manual assembly. *Manufacturing systems and technologies for the new frontier* (pp. 245–250). s.l.: Springer Verlag London Limited. https://doi.org/10.1007/978-1-84800-267-8_50

Strang, D. (2016). Kommunikationsgesteuerte cyber-physische Montagemodelle. Dissertation, Technische Universität Darmstadt, Darmstadt. ISBN 978-3-8440-4594-9.

Strang, D., & Anderl, R. (2014). Assembly process driven component data model in cyber-physical production systems. In *Proceedings of the World Congress on Engineering and Computer Science* (Vol. 2). ISBN 978-988-19253-7-4.

Ullrich, G. (1995). Wirtschaftliches Anlernen in der Serienmontage—Ein Beitrag zur Lernkurventheorie. Dissertation, Gerhard-Mercator-Universität Duisburg, Duisburg. ISBN 978-3-8265-0868-4.

Verein Deutscher Ingenieure. (2019). VDI 2221: Design of technical products and systems—Model of product design.

Verein Deutscher Ingenieure. (2020). VDI/VDE 2206: Development of cyber-physical mechatronic systems (CPMS).

Wilkowska, W., & Ziefle, M. (2013). User diversity as a challenge for the integration of medical technology into future smart home environments (pp 553–582). https://doi.org/10.4018/978-1-4666-2770-3.ch028

Wright, T. P. (1936). Factors affecting the cost of airplanes. *Journal of the Aeronautical Sciences, 3*(4), 122–128. https://doi.org/10.2514/8.155

The Effects of the Digital Twin of Humans

From Computer-Assisted Work to the Digital Twins of Humans: Risks and Opportunities for Social Integration in the Workplace

Sarah Brunsmeier, Martin Diewald, and Mareike Reimann

Abstract In digitalised workplaces, a multitude of data is generated via computer-assisted work (CAW). These data enable comprehensive digital representations of employees, also referred to as the digital twins of humans. The possible consequences of CAW for the social integration of employees in the workplace have hardly been studied. With our contribution, we want to gain a detailed picture of how the design, implementation, and individual appropriation of CAW can shape working relationships, and we want to determine what role human digital twins can play in social integration in the workplace. We distinguish between CAW for information and communication, which directly shape relationships, and CAW that has the potential to influence relationships indirectly through other workplace characteristics, including control, routine work, autonomy, privacy, datafication, and transparency. The current literature shows that risks like isolation and bullying outweigh opportunities like extended cooperation possibilities. Furthermore, merging CAW into human digital twins could increasingly jeopardise social relationships. However, there appear to be design, implementation, and individual appropriation opportunities that could allow CAW to prevent threats to social integration in the workplace or even to foster positive relationships.

Keywords Workplace relationships · Social integration · Social networks · Computer-assisted work · Computer-supported work · Digital assistance systems · Digital twins of humans

S. Brunsmeier (✉) · M. Diewald · M. Reimann
Faculty of Sociology, University Bielefeld, Universitätsstraße 25, 33615 Bielefeld, Germany
e-mail: sarah.brunsmeier@uni-bielefeld.de

M. Diewald
e-mail: martin.diewald@uni-bielefeld.de

M. Reimann
e-mail: mareike.reimann@uni-bielefeld.de

© The Author(s), under exclusive license to Springer Nature Switzerland AG 2023
I. Gräßler et al. (eds.), *The Digital Twin of Humans*,
https://doi.org/10.1007/978-3-031-26104-6_10

1 Introduction

The potential that digital systems have to alter our social networks for better or worse is mainly investigated and discussed for big information and communication tools like Facebook or Instagram. The possible consequences of computer-assisted work (CAW) have not been studied as thoroughly. CAW can shape the embeddedness of employees in social networks with colleagues and superiors in multiple ways. Social relationships in the workplace are in the first place professional, i.e., they facilitate work processes and fulfil work tasks within a hierarchical position structure. However, this requires relationships to be not only functional but also personal to some degree. There is considerable variety in the degree to which employers and employees emphasise the personal side of social relationships in the workplace. However, in all cases, workplace contacts can foster a sense of belonging and provide useful resources such as help with tasks, emotional support, opportunities for advancement, personal growth, and increased well-being (Colbert et al., 2016), even beyond a work-related context. Workplace relationships are an important part of the overall social network of employees. However, these contacts may also be shaped by unwanted criticism and harassment, as well as bullying. Bullying in the workplace is a common problem (Lange et al., 2019) with many negative consequences for employees and companies (Bonde et al., 2016; Einarsen & Mikkelsen, 2003; Einarsen & Nielsen, 2015; Salin & Hoel, 2003; Mikkelsen & Einarsen, 2002).

In this chapter, we propose an approach that differentiates between the different purposes for which CAW is designed. This will help us to better understand how CAW may shape workplace relationships in both directions, i.e., whether it provides opportunities and/or has risks.

So far, CAW has been primarily discussed in the context of digital assistance systems (DASs) in Industry 4.0, with a focus on the shop floor. However, CAW has also been utilised in the office for a long time. We define CAW as work supported by digital technologies through hardware and software. CAW includes computer-supported cooperative work (CSCW) and technologies that are not based on cooperation, i.e., they assist single employees. Computer-assisted technologies are technical aids that are meant to reinforce the working power and work performance of employees, either physically or mentally (Apt et al., 2018). The aim is to accelerate and compress work processes (Niehaus, 2017). On the shop floor, computer-assisted technologies like DASs are often used to form the interface between humans and machines that are connected via the internet, and they are particularly used to increase the amount of information about work processes and manage the increasing amounts of process-produced data (Niehaus, 2017). DASs in the form of mobile devices in logistics, which enable an overview of the manufacturing inventory at any place and at any time, are an example of CAW. These mobile devices often track the work steps of employees. However, common computer software programs such as Microsoft Office and conference and e-mail programs, which are primarily used in an office context, are also examples of CAW.

We consider three perspectives on how CAW functions. *First*, CAW is designed to serve specific functions in the work process. *Second*, CAW forms a socio-technical system (Hirsch-Kreinsen & ten Hompel, 2017). Socio-technical systems consist of technological, organisational, and human subsystems that reciprocally shape each other (Hirsch-Kreinsen & ten Hompel, 2017). The perceptions, thoughts, and actions of employees are influenced by the design and implementation of CAW. In turn, the design and implementation of these technological systems can affect the perceptions, thoughts, and actions of employees. *Third*, data collection by very comprehensive and/or several digital tools in the same workplace leads to a new quality of digitalised workplaces, which we capture using the concept of the digital twins of humans. The significance of the digital twins of humans goes beyond the significance of single, focused digital technologies. We will now specify how these three perspectives are related to social integration in the workplace.

1. **Types of CAW**: When considering the question of how CAW specifically shapes workplace relationships, it is important to distinguish between two different types of CAW. Type 1 CAW is intentionally designed to support information and communication for those involved in a value chain. Type 2 CAW is designed for purposes other than communication, e.g., information storage or planning and scheduling work processes. For type 1 CAW, an influence on social relationships is evident from the very beginning, since it directly shapes how colleagues and superiors communicate with each other. This applies to the sheer amount of contact, as well as the quality of these relationships and their functions. For type 2 CAW, the influence on social relationships is less obvious. Type 2 CAW initially directly influences workplace characteristics such as work autonomy, control, monitoring, or performance measurement. However, type 2 CAW may shape workplace relationships indirectly. This is because perceptions of being controlled, monitored, or incapacitated by type 2 CAW shape what employees expect from colleagues or superiors and to whom they assign responsibilities for work tasks.

2. **Sources of influence of CAW**: Any such consequences of type 1 and type 2 CAW for social relationships are not just the deterministic outcomes of a technological system. Considering CAW as a socio-technical system, according to the second perspective of our approach, means that at least three sources of influences of CAW on workplace characteristics can be distinguished: (1) the *technical design* of CAW, i.e., its technical specifications and configurations; (2) the *implementation* of CAW in the organisational environment, i.e., how alternative configurations are used to adapt CAW to the specific organisational aims, structures, and culture; and (3) the *individual appropriation* of CAW when digital technology is used according to each employee's own goals and needs. The technical design of CAW includes, for example, various technical functions that provide a framework for how CAW influences workplace characteristics. For type 1 CAW, one such function could be communication via chat messages. The technical design of CAW determines the ways in which it can be implemented by employers and used by employees. The technical design usually offers leeway in implementation

in a work context. In the case of chat communication, employers could determine the communication occasion and communication partner through instruction or through administrative technical settings. The designs and implementation limit the individual appropriation of CAW. Nevertheless, there are often different design options for employees, such as, for example, the determination of the length and frequency of messages in the exemplary chat itself. The individual appropriation of this system is likely to affect workplace relationships. The three levels of influence of CAW provide information that can be used to create CAW in a way that allows the intended influences on workplace relationships to be achieved and unwanted influences to be avoided. Technical design, implementation, and individual appropriation can therefore affect workplace relationships directly or indirectly through other workplace characteristics. Individual perceptions and evaluations of the functions and implementation of CAW, especially with respect to other workplace characteristics such as autonomy, control, and performance measurement, are decisive mediators and moderators of the possible consequences of such CAW for workplace relationships.

3. **Digital twins of humans**: Almost all CAW produces immense amounts of data not only about the functioning of the utilised technologies but also about the users. The richness of these data gains a new quality if data from CAW with various purposes and fields of activity, or data from very comprehensive CAW with different fields of activity, are linked. Data linkage allows the creation of comprehensive representations of employees, which can be called the *digital twins of humans* (Gräßler & Pöhler, 2020). Whereas a linkage of within-firm process-produced data with other personalised data (e.g., medical records) is very unlikely to be realised due to strict data protection rules in Germany, a linkage of all within-firm data is much more likely to be achieved and harder for the state to control. Human digital twins could be used, for example, to further advance speed in the production process by mimicking employees by linking human digital twins directly to machines that are connected via the internet (Gräßler & Poehler, 2017). Gräßler and Poehler (2017) described the example of a digital twin of an employee linked to a production system that could independently intervene in the production system based on the stored preferences of the employee. In this case, the employee would monitor the procedure, for instance, via a mobile device, but usually would not intervene.

In the following, we consider the three aforementioned perspectives to demonstrate how the consequences of CAW for workplace relationships can be traced back to specific components of CAW within the framework of socio-technical systems and human digital twins. A review of the existing literature reveals what we know and what we still do not know about recognising the risks and possibilities of the growing use of CAW, and what mechanisms can be used to avoid risks and enhance opportunities. In the next section, we review the existing evidence concerning the consequences of type 1 CAW. In Sect. 3, we do the same for type 2 CAW. We conclude Sect. 3 with a discussion of home-based telework as a new form of work in which mainly, but not only, type 1 CAW comes into play. Section 4 discusses to what extent the digital twins of humans are currently legally conceivable and how data linkages across several technological CAW systems may exert an additional influence on workplace relationships over and above the influence of single technological systems. A final section summarises the existing evidence on risks and opportunities and discusses directions for further research, as well as strategies for coping with risks.

2 Direct and Indirect Consequences of Type 1 CAW for Workplace Relationships

Type 1 CAW (CAW for information and communication) has become an almost ubiquitous and self-evident part of the workplace environment. Widespread examples are software programs like Microsoft Teams with many areas of application, or the use of more specialised software like e-mail and, largely due to the COVID-19 pandemic, conference tools like Zoom. According to Niehaus (2017), by implementing digital technologies with certain functions, employers are usually trying to achieve an increase in efficiency and effectiveness. Employees are confronted with different functions if a single digital technology has various functions or if employees use many digital technologies at the same time (Niehaus, 2017). Thus, various factors that influence social relationships can be intermingled in the everyday work of employees. The way in which CAW exists within social networks partly depends on the design prerequisites, the implementation, and the individual appropriation of different functions. Ultimately, however, what is especially relevant for individuals is how they perceive the design and implementation of CAW, e.g., whether they perceive it as transparent, an invasion of privacy, challenging their autonomy, or alienating. In the following, we identify mechanisms that may impact social relationships in the context of information and communication functions. As information and communication systems can each directly impact relationships, we consider them separately.

Information Systems
Companies increasingly use digital networks to enable time- and location-independent information exchange to increase efficiency and innovation (Trier & Richter, 2015). Via access to information pools, intellectual knowledge can be made easily accessible to others, and information can be reproduced and further developed (Boes et al., 2020). Digital technologies for type 1 CAW provide information, recommendations for action, or instructions aimed at structuring communication and cooperation processes in the company (Niehaus, 2017). These information functions can replace human communication and interaction to some extent. For example, information systems can automatically provide information for certain work steps for the training or further education of employees (Apt et al., 2018).

How the design of such tools affects work organisations ultimately depends on the way these tools are implemented in the context of each organisation. The emerging networks are complex (Trier & Bobrik, 2009) since inclusion and exclusion from participation may not completely overlap for several information circles. This can cause irritation and a perceived lack of transparency. Through the individual appropriation of the functions of digital technologies, such as e-mail distribution lists, CAW can also be used in a directly inclusive or exclusionary way. To some extent, the distribution of information can take place at the employee's own discretion and thus can have an impact on the social integration of employees being included in or excluded from these lists.

A major issue that is typically discussed for type 1 CAW systems used for information distribution is the risk of information or work overload. Due to the implementation method of the organisation and depending on the individual appropriation of the system by employees, overload often results in negative impacts on social relationships. For example, e-mails can be received and answered at any time (Salin & Hoel, 2003). Work cannot be done fast enough, which results in employees being overloaded— 'the feeling of having too much to do in too little time' (Kelly & Moen, 2020). The case study by Kelly and Moen (2020) indicates that type 1 CAW is not the root cause of overwork, but it helps employees meet the always-on expectations of their superiors. The study shows that both individuals and intra-departmental teams often think of themselves first and foremost when working in an environment that is characterised by overload and insecurity; this is made possible by the advancing technologies used for type 1 CAW. In this case study, this lone wolf mentality negatively impacts social relationships and social support. The authors concluded that the negative effects can be partially reversed if type 1 CAW is implemented in an organisational environment in which flexibility is not expected from employees but is given to employees. Control over their work location and time, a focus on results rather than attendance, more online work, fewer low-value tasks, and social support provide employees with more time to help others and improve their interactions with their colleagues.

Communication Systems

According to Boes et al. (2020), the internet has created both a new 'information space' and a new 'social action space'. In addition to the information function, people can communicate and collaborate in different ways and relate to each other. In a workplace context, digital technologies with communication functions make it possible to use these new spaces at any place and at any time. Collaboration is also increasingly taking place digitally. New forms of collaboration and knowledge exchange are emerging, and work is being adapted to information flows (Boes et al., 2020). The internet provides the opportunity to build social relationships and communities. Users want to connect and communicate—especially with like-minded people (Trier & Bobrik, 2009).

According to a study of a digital organisational network by Trier and Richter (2015), organisational digital networks have the potential to promote the division of labour; structuring takes place via subject areas for which there are experts. For example, links can be created between employees who work at different company locations, even across countries. This does not create strong reciprocal relationships;

instead, it creates a dense network with weak links (Trier & Richter, 2015). Digital social networks therefore stand in the way of greater social integration, which is often made possible by personal relationships. Intra-company networks can thus be weakened by replacing face-to-face encounters with weak relationships within type 1 CAW. Trier and Richter (2015) identified few active participants who want to advance an issue in digital networks and many passive recipients of information. Employees assign themselves to different topics and strive for efficiency in obtaining information and support in solving problems. Weak relationships sometimes result in new modes of interaction, such as collaboration on projects (Trier & Richter, 2015). Digital communication functions can thus have a positive impact on relationships by fostering support and enabling new contacts that allow a sense of belonging because of similar interests or working topics. In this respect, the direct communication functions of digital technologies are an opportunity for improved social integration in the workplace.

However, access to digital communication systems in the workplace is rather restricted. The organisations that implement the communication systems often also determine who is allowed to use them. Management can enable contacts but can also restrict contacts by excluding employees from systems (Trier & Bobrik, 2009). Furthermore, digital communications between employees can be stored. The stored communications sometimes become content in digital networks that provide information to others who did not participate in the communication (Trier & Bobrik, 2009). Moreover, the ability to control the progress of users through data storage using digital communication systems increases the transparency of work (Boes, 2016). If the communication and cooperation processes are restrictive due to the way that employers implement the digital technologies of type 1 CAW, employees' scope of action could decrease (Niehaus, 2017). A limitation of spontaneous interaction dynamics beyond predetermined procedures could be the result.

The individual appropriation of digital communication can also be a risk for informal relationships in the workplace. The changed possibilities of cooperation via digital networks and the joint processing of data can limit personal social relationships. Informal encounters might diminish due to the use of digital communication media (Goll, 2008). An increased anonymity through digital communication can eventually lead to the increased isolation of employees (Cooper & Kurland, 2002). Overall, digital networking and communication functions can increase the number of working relationships and the feeling of belonging, but these digital contacts cannot replace personal encounters. Employees must take care of these personal contacts somewhat independently. Moreover, the inappropriate use of digital communications functions can have negative consequences for workplace relationships. For example, when sensitive issues are communicated digitally, tensions can easily arise between communicators because written words often leave a lot of room for interpretation, and misunderstandings cannot be cleared up immediately.

Some employees cannot cope well with communication overload and the acceleration of work processes, regardless of the possible flexibility of the system (Kelly & Moen, 2020). The implementation of communication systems that are meant to create constantly accessible and multitasking employees fundamentally poses a risk to workplace relationships. The question of how employees deal with increased degrees of freedom arises. Ultimately, it often comes down to the ability of employees to set boundaries and create structure for themselves (Rexroth et al., 2014). According to Kelly and Moen (2020), stress arises among employees because expectations often cannot be met and private life and work life intermingle. To cope with the increased time pressure, many employees use multitasking and attend, for example, online meetings while performing other work processes. This can jeopardise relationships in the workplace. Communication overload and the desire to meet demands can mean that employees have less time for mutual support and attention. This leads to a bad atmosphere and permanent tensions within and across teams to the point that team leaders will instruct their subordinates not to help out other teams.

3 Indirect Consequences of Type 2 CAW for Workplace Relationships Through the Shaping of Other Workplace Characteristics

Type 2 CAW directly shapes other work characteristics that might indirectly influence workplace relationships. It is important to distinguish between functions that control employees and have the potential to decrease their scope of action, and those functions that support employees and may enhance their scope of action (Niehaus, 2017). A common function of type 2 CAW is to monitor work processes, contexts, and employees using sensors (Apt et al., 2018). DASs such as tablets, smart watches, and data glasses generate and permanently store information on the location, speed, and quality of employees' work. Some DASs for type 2 CAW use highly intelligent algorithms that can recognise patterns, interpret language, and incorporate employees' feelings and motivations (Apt et al., 2018). In addition to monitoring, restrictive requirements on work content also serve to control employees and are often used for performance assessments (Niehaus, 2017). Some DASs assess the performance of employees independently (Apt et al., 2018). These evaluations can be included in general performance appraisals from supervisors (Gal et al., 2020). However, feedback from DASs that employees themselves have access to can also be used to support employees in their learning processes (Apt et al., 2018; Gräßler et al., 2020). Through a learning function, qualification times can be reduced and employees can be deployed more flexibly, for example, through job rotation (Gräßler et al., 2021; Niehaus, 2017). Another way to support employees through DASs is to point out alternatives to help with decision-making and therefore enhance employees' scope of action (Niehaus, 2017). DASs can also be used to promote relief for employees in terms of ergonomics and security at the workplace (Niehaus, 2017).

It is unclear how the functions that constitute design aspects are perceived by employees. Do employees feel sufficiently involved in the process of data collection and storage, or do they not understand the data storage process in detail? Do they feel that their privacy is being restricted? These questions arise regardless of whether technologies of type 2 CAW are implemented for the purpose of control or for the purpose of support. It is also crucial to consider how comprehensive the data storage is and what framework conditions are provided by the employer to track data storage and processing (Gal et al., 2020; Hansen & Thiel, 2012).

In the following, we distinguish between the individual level and the organisational level. Changing workplace characteristics on the individual level like control, routine work, autonomy, privacy, datafication, and transparency in the context of type 2 CAW are taken into account. On the organisational level, tensions due to fears and worries and solidarity caused by the similar experiences of employees are considered.

Work Characteristics Like Control, Routine Work, Autonomy, Privacy, Datafication, and Transparency

The handling of the storage of data concerning employees' work via CAW can influence workplace relationships indirectly through workplace characteristics such as the perception of control, routine work, autonomy, privacy, datafication, and transparency. All these characteristics have been associated with workplace relationships. Type 2 CAW can increase the importance and intensity of these characteristics using various functions. This is why relationships in the workplace can also develop in certain directions in an amplified way. In the following, the influence of type 2 CAW on these characteristics, as well as its effects on social relationships, are presented.

Control

Data storage can be a control medium that is used to closely monitor employees (Niehaus, 2017), but it can be used for performance evaluation at the same time (Moore, 2018). Performance evaluations via automatic data storage might involve a lack of detailed feedback for employees (Gal et al., 2020). For example, when DASs are introduced to prevent errors and improve performance, performance may not be discussed with employees, but errors and detailed information on employees' work processes are collected (Moore, 2018). However, according to Schaupp (2022), algorithmic control does not necessarily mean that supervisors check every work step that an employee performs. Supervisors sometimes use DASs that provide automatic feedback to employees about their work steps to create the illusion of control, even though they themselves do not control everything (Schaupp, 2022). Whether the feedback from digital technologies is acted upon by employees often depends on how strongly they worry about losing their job (Schaupp, 2022). According to Wood (2020), low job security in an organisation combined with a large amount of technological and human surveillance leads to the excessive power of supervisors over their subordinates. Supervisors can use CAW to monitor not only performance but also very personal aspects of employees, such as their behaviour, thoughts, and feelings (Ball, 2021). The power this gives them is often abused and used to harass and bully subordinates (Wood, 2020). Coercive control can encourage the humiliation of employees by management and hostility towards those who exercise control (Crowley, 2014). The exercise of coercive control in itself can negatively affect the dignity of employees (Crowley, 2014), which can undermine solidarity, knowledge sharing, and friendships between colleagues (Hodson, 1996).

Routine Work

The control of employees, for example, through routine tasks in the context of new digital technologies, can be seen as a form of coercive control (Melzer & Diewald, 2020). Furthermore, frustration due to a lack of opportunities to monitor and control one's work, unclear and conflicting goals, and a lack of constructive leadership are consequences of monotonous work (Einarsen et al., 1994). These aspects of routine work often result in increased bullying (Einarsen et al., 1994). Since routine work can be a form of coercive control, the negative effects on relationships described in the paragraph above might also be a consequence of routine work.

Autonomy

The autonomy that employees have or do not have may also be influenced by type 2 CAW (see Chapter "Work Autonomy and Adaptive Digital Assistance in Flexible Working Environments" by Gensler et al. on work autonomy and DASs). A function that provides employees with real-time feedback based on software-generated data may enhance employees' autonomy by giving them the opportunity to learn independently (Apt et al., 2018). The high level of digital surveillance made possible by DASs instead leads to increased stress (Aiello & Kolb, 1995; Castanheira & Chambel, 2010; Khanchel, 2020) and the perception of less autonomy (Kensbock & Stöckmann, 2021) for employees. More stress and less autonomy are linked to increased bullying at work (Einarsen et al., 1994; Salin & Hoel, 2003). Depending on how certain functions of DASs affect autonomy, both an increase and a decrease in bullying at the workplace are therefore conceivable.

Privacy

Moreover, if data that employees perceive to be irrelevant to the work task are stored, and overall excessive monitoring takes place or is at least feared, data storage can be perceived as an invasion of employees' privacy (Ciocchetti, 2011). This is typically the case for employees working from home: Any surveillance intrudes into the private sphere of life and the boundaries between work and family are blurred. In this case, some employees are afraid to use the monitored electronic devices (Ciocchetti, 2011). According to Moussa (2015), electronic monitoring could lead to perceptions of increased stress among employees and consequently to lower satisfaction, loyalty, motivation, and integrity. This stress might increase pressure and tension among employees (Moussa, 2015), and thus, it may increase negative social relationships between employees.

Datafication

The surveillance of employees is often linked to performance evaluations (Moore, 2018). Whether performance appraisals have negative effects on employees seems to depend heavily on the context of their use. Performance appraisals that involve the automatic storage of data can also be used for the purpose of surveillance. As stated by Gal et al. (2020), datafication carries the risk that employees are seen only as the data that are collected about them. The focus on data and data-based performance evaluations could lead to a lack of understanding of one's own actions and their effects by employees (Gal et al., 2020). Gal et al. (2020) also suggested that communication could suffer, resulting in more difficult socialisation within a company.

Transparency

Performance evaluations are not necessarily transparent for employees or supervisors (Gal et al., 2020). This can negatively affect data transparency, i.e., one is not able to see the stored data concerning oneself, as well as decision transparency, i.e., automatic data collection and processing do not lead to objective and transparent decisions about employees (Bannister & Connolly, 2011). Sometimes, even supervisors are not able to understand the ways in which DASs process data and which facts the output is based on (Gal et al., 2020). Nevertheless, they still use this output to make decisions about employees' work; therefore, decisions are often not traceable and employees feel powerless (Gal et al., 2020). Furthermore, by collecting detailed data on employees, their behaviour can be manipulated without them being aware of the manipulation (Gal et al., 2020). In contrast, performance appraisals that allow employees to make decisions and become more efficient, for example, through real-time feedback on their learning progress, are not necessarily linked to surveillance (Apt et al., 2018). This indicates that performance evaluations can be a part of coercive control, negatively affecting relationships, but they do not have to be. Regardless, a lack of transparency in performance appraisals and constant surveillance could in itself be perceived as harassment (Moore, 2018). Whether data retention has negative consequences for employees depends in particular on how monitoring is implemented in the organisation (Ball, 2010). Data protection and the transparency of the purpose of the data can make it possible to avoid many problems (Funk et al., 2019; Yerby, 2013). The perception of transparency, intervenability, and data sparseness protect privacy and increase the acceptance of cyber-physical systems (Hansen & Thiel, 2012). The perception of employees regarding workplace characteristics like routine work, autonomy, control, privacy, datafication, and transparency seems to be important for the arrangement of workplace relationships. However, this is rarely addressed in the relevant literature.

There is research on social integration that looks at digitalisation as an overall construct. According to a representative study by Melzer and Diewald (2020), in addition to the many indirect negative effects on social integration of some DAS functions, working with DASs in general can indirectly protect all employees from bullying by supervisors. The authors assume that digital work could change the relationship between supervisors and subordinates. Highly skilled employees may be more protected from bullying by individual digital competencies that are important for the company. The study indicates differences between employees with different qualification levels. For example, it shows that digitalised work can protect low-skilled employees from bullying by colleagues. Low-skilled employees may be less likely to experience bullying from colleagues either because the automation of tasks reduces interactions or because solidarity develops among employees who must deal with the newly introduced systems (Melzer & Diewald, 2020).

Contextual Influences of the Workplace

In this subsection, we address the consequences that predominantly type 2 CAW can have for larger social networks in organisations. The focus on social networks continues to provide information about individual relationships, as social networks consist of these relationships. Even employees who do not work or work only a little with digital technologies can be influenced by the implementation of CAW in the work organisation (Melzer & Diewald, 2020). The introduction of CAW can be associated with concerns and fears among employees (Heath et al., 1993) related to performance restrictions and job loss (Green, 2011; de Witte, 1999). Responsibilities in relation to newly implemented digital technologies need to be negotiated (Baillien & de Witte, 2009), which can lead to conflicts and competition for newly created key positions (Vallas, 2006). From the management side, the implementation of CAW can lead to more authoritarian management styles when there is resistance from employees (Salin & Hoel, 2003). Melzer and Diewald (2020) found that highly skilled employees in workplaces with higher shares of employees working with technological systems experience more bullying by supervisors. They explain this finding using the fact that the requirements of the new system lead to more conflicts between highly qualified employees and their superiors because the power differences are smaller. In their opinion, it is also possible that highly qualified employees perceive new authoritarian management concepts as bullying because they are not used to this kind of interaction.

However, for all groups of employees, and especially for the low-skilled employees, the shared experience of digital control can lead to an increased solidarity with colleagues (Schaupp, 2022; Wood, 2020). Employees performing standardised work processes with algorithmic control across different countries and companies often feel solidarity with each other because their work processes, in the form of carrying out automatic instructions, are becoming more and more similar (Schaupp, 2022). In general, insecurity and surveillance can motivate people to organise independently or collectively in unions, as shown by Wood (2020). In his case study of a retailer, social media was used for open resistance to successfully diminish the employer's control. Through social media, workers from different branches of the retailer were able to share similar experiences and support each other emotionally. With the help of the network, they were able to reveal that control was not isolated but structurally implemented in all shops. These findings contradict those of Zuboff (2019), who stated that algorithmic control and solidarity are usually not compatible with each other. The literature suggests that when implementing CAW, the possible concerns and fears of employees should be taken seriously. If management is transparent, tensions among employees may be avoided. If many employees are affected by negative workplace characteristics, this can create solidarity among them.

Pathways of Influence on Relationships Across Different CAW Types

CAW impacts workplace relationships initially in different ways depending on its technical functions. Figure 1 illustrates the different pathways of influence on relationships in the workplace according to the two CAW types. Type 1 CAW mainly has a direct impact on the individual relationships of employees through the features of digital communication. Type 2 CAW exercises an indirect impact on working relations through other workplace characteristics. In the latter case, a distinction must be made between effects at the individual level and effects at the organisational level. The organisational level often also affects employees who do not work

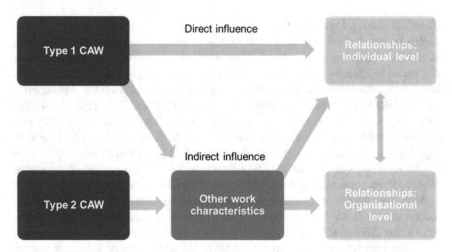

Fig. 1 Pathways of influence on relationships for different types of CAW (illustration by the authors)

with digital technologies. Previous studies on the organisational level mainly address type 2 CAW. However, there are also indications that type 1 CAW indirectly affects organisational-level relationships in rare cases, for example, when the CAW systems are newly implemented. Type 1 CAW could also have an indirect impact at the individual level, as, for example, communications can be monitored, and surveillance in the form of coercive control might affect relationships negatively (Crowley, 2014). Finally, it is important to consider that relationships at the individual and organisational level are interrelated.

Excursion: Home-Based Telework as an Example of Different Digital Technologies for CAW
Because of the worldwide COVID-19 pandemic, working from home has been the most-discussed form of digitally enabled work in 2020 and 2021. For employees working from home, different digital technologies for CAW come into play that determine whether CAW has positive or negative consequences for employees and also for social relationships. CAW may lead to various dilemmas for employees and companies (Diewald & Nebe, 2020).

Collaboration and Communication
For the most part, type 1 CAW is used to enable employees to relocate their work from the workplace to their homes. For example, via VPN technologies, employees are provided with access to internal information and data. Moreover, previous face-to-face communication with colleagues and supervisors is heavily replaced by digital forms of communication, such as e-mails or communication tools implemented in information platforms. As a consequence, the previously discussed benefits and risks of type 1 CAW are even more present when employees are working from home. Most importantly, type 1 CAW alters interaction processes in work relationships

(Dambrin, 2004). Direct feedback options are reduced and lively team work processes are made difficult (Knights & McCabe, 2003). Especially when the focus is on individual work performance at home, information flows, creative processes, and social integration in the company may be impaired (Diewald & Nebe, 2020). Electronic communication makes employees less visible to colleagues and supervisors, resulting in increased anonymity (Cooper & Kurland, 2002). This may lead to social and professional isolation (Collins et al., 2016; Cooper & Kurland, 2002), and employees tend to socially distance themselves from other employees because they experience less cohesiveness when nobody is around (Cohen & Bailey, 1997). This is why employers as well as employees seem to favour arrangements that involve working from home and working at the workplace (Kunze et al., 2020).

Moreover, the risk of stress due to an increasing dissolution of boundaries between work and private life has increased significantly due to mobile work and working from home (Michel & Wöhrmann, 2018). The permanent presence of technological tools that connect employees to CAW makes the separation of work and private life more difficult, and permanent availability for work-related communication, even outside working hours, becomes more likely (Boswell & Olson-Buchanan, 2007). Increased stress, in turn, increases the likelihood of unsupportive social relationships and bullying (Salin & Hoel, 2003).

Work Characteristics Like Autonomy, Stress, and Routine Work

In addition, other aspects of CAW and its positive and negative consequences for social relationships are likely to be exacerbated when employees are working from home. Because communication and the use of data through CAW is often limited to predefined rules and procedures, individual autonomy in using these digital technologies for CAW may be lower. On the one hand, CAW for home-based telework may involve more routine tasks that can be performed easily with digital tools at home, without the need for complex interactions with colleagues, customers, or work equipment at the workplace. On the other hand, routine tasks can be performed more easily because they already have been implemented in a digital work environment and thus, they can be performed more effectively than they could be without CAW. Moreover, access to the work of colleagues and supervisors, for example, in the form of jointly edited documents, can also help to avoid unnecessary duplicate work or to improve the workflow, because the ability to work asynchronously means that employees do not have to wait for other participants in the work process to be available. CAW can also be used to monitor employees working from home. In this case, Diewald and Nebe (2020) stated that there can be a conflict between controlling the work of employees and preserving the privacy of employees, which is particularly evident when employees are working from home. The presence of employees at the workplace allows the monitoring of work that could not be monitored at the employee's home without violating their privacy. Video surveillance in one's own home, for example, would not be compatible with the protection of privacy. The employer's desire for control and the privacy of one's own home are opposed when it comes to data collection and storage (Diewald & Nebe, 2020). As the feeling of an invasion of privacy can lead to increased tension between individuals (Moussa, 2015), attempts at control by superiors could negatively affect the relationship between the parties involved.

4 From CAW to the Digital Twins of Humans

CAW can be linked to human digital twins through the increasing storage of employee data. In the following, we explain which changes in workplace relationships could occur in the context of human digital twins, with reference to the mechanisms that influence social relationships in the context of CAW. However, possible future scenarios of human digital twins, including specific benefits and risks for social relationships, may differ from current practical and legal possibilities. Therefore, we first look at the extent to which human digital twins are legally conceivable at the moment.

Legal Framework for Storing Employee Data within CAW

From a legal point of view, German law defines limits regarding collecting and processing employee data. Biometric and genetic data may not be used to identify individuals, and total surveillance is generally prohibited (Kort, 2018). The storage of data for the performance evaluation of employees is permitted for the purposes of personnel planning, for the appropriate deployment of employees, and if it is important for the professional careers of employees, for example, with regard to professional development opportunities and the comparison of employees (Franzen, 2021).

However, there are still major gaps in the law when it comes to dealing with big data. The handling of large datasets, which are analysed by algorithms, among other things, is addressed neither in the General Data Protection Regulation (GDPR) nor in the Bundesdatenschutzgesetz (BDSG) (Kort, 2018). In Article 22 of the GDPR, there is only a prohibition of automated decisions in individual cases; this, however, does not apply if the data subjects have given their consent. Furthermore, unless the consent of the employee is given, the collection and processing of employee data in general is only permitted if the data processing is proportionate, i.e., the interests of the employer must be weighed against the interests of the employee (Kort, 2018). The appropriateness and proportionality of video surveillance in the workplace is measured, for example, according to how many persons are being monitored, whether surveillance is carried out openly or covertly, why surveillance is being carried out, and how long surveillance is being carried out (Kort, 2018).

According to the legal requirements, human digital twins that perform parts of human tasks are conceivable. However, due to the prohibition of the total surveillance of employees and the right to informational self-determination, human digital twins can possibly only be used for specific parts of the work process and might often be incompatible with the privacy requirements of working from home. Ultimately, however, the question of how organisations will actually deal with legal frameworks also remains. Despite the legal requirements, it seems that surveillance in Germany is not always transparent and proportionate (Staab & Geschke, 2020). It is also necessary to distinguish between the legal possibilities of implementation and concerns and fears among employees. Fears concerning certain impacts that may be associated with the way that the digital twins of humans are implemented may also be present when certain methods of implementation are not legally allowed. Therefore, it is crucial to include employees' perceptions when assessing and studying these impacts.

The Digital Twins of Humans and Possible Changes in Workplace Relationships
The implementation of human digital twins could exacerbate the identified risks for workplace relationships due to CAW. The digital twins of humans require immense monitoring and the storage of employee data (Berisha-Gawlowski et al., 2021). Because data are linked, the risks described in the literature can occur. For example, the use of digital twins could replace a large part of current coordination processes if the interfaces between people and machines become partly automatic. Currently, much coordination takes place via the communication and networking capabilities of type 1 CAW. These coordination processes between employees could be reduced if the digital twins of the employees carry out some of the coordination automatically. This could result in fewer opportunities for digital communication and networking. An exacerbation of isolation could be a consequence of this scenario.

The evolution of CAW into human digital twins could also aggravate information overload, as human digital twins are another means of displaying information about stored data. At the same time, it must be asked whether and how human digital twins, proportionally controlled by the humans they represent, are in contact with each other and what this means for the social relationships of the employees. How does one human digital twin affect another human digital twin, and what control options do employees have? With regard to the implementation of human digital twins, the question of how they are integrated into the work processes of employees arises. They could be used specifically to support employees so that employees have more creative time to design their projects (Nöhammer & Stichlberger, 2019). If human digital twins are used to lower the workload of employees, further negative effects of information overload could possibly be avoided.

Unfortunately, based on the current state of the literature, the risks outweigh the opportunities for positive change. CAW has the potential to reduce routine tasks by using more complex task profiles, but it can also increase routine tasks due to restrictive instructions (Niehaus, 2017). However, the digital twins of humans in particular seem to increase the amount of monotonous work. For example, there is the possibility that employees' work will become more monotonous, as human digital twins could take over many formerly complex human tasks like decision-making. The human tasks would then be limited to controlling the digital twins (Gräßler & Poehler, 2017). In a scenario with more monotonous labour, relationships could be affected negatively. Human digital twins could also be a new medium that is used to control and monitor employees even more than is possible with CAW alone. If privacy is violated in the context of the data storage of human digital twins and autonomy is reduced by coercive control, bullying in the workplace could increase. If the digital twins of humans exist, it is possible that employees in particular will be viewed and evaluated based on their digital representations. This can influence socialisation in the company when relationships are based mainly on data and not on human communication (Gal et al., 2020). Moreover, due to type 2 CAW with automatic instructions, it might not even be necessary to be trained by other employees (Gal et al., 2020). Evaluations often take place using the stored data of employees (Moore, 2018). Therefore, there is a serious risk that management will use human digital twins to exercise coercive control and to treat employees like numbers. In addition to an increased risk of bullying due to coercive control (Crowley, 2014), favouring the data over the human being could inhibit intrinsic motivation in employees and create conflict with supervisors. Furthermore, if certain digital information about employees is considered objective, this poses risks to relationships. Information is based on employees' work steps, but the digital technologies of CAW can only process this information according to their design and implementation. Due to the amounts of data required for human digital twins, it is likely to be a challenge to carry out monitoring and data storage appropriately, transparently, and with a human's active participation.

Furthermore, the introduction of human digital twins will cause organisational changes that could affect all employees. Therefore, this new work situation could create worries and fears among employees, necessitate the renegotiation of roles, and thus increase tensions between employees and supervisors. If the risks to relationships are not considered in the design and implementation of CAW, the amalgamation of data on employees into digital twins could result in interpersonal tensions and the isolation of employees.

5 Discussion

In commercial fields, such as the film industry and literature, relationships have received great attention in the context of advancing CAW. In academia, the topic has only been considered marginally. However, a detailed look into the current research shows that the issue requires greater attention in academia. According to our findings, CAW can have consequences for workplace relationships through its design, implementation, and appropriation by employees. The consequences for relationships due to CAW could be intensified by the connection of stored data about employees to human digital twins. We were able to identify some risks but also opportunities for social integration in the workplace in the context of CAW. However, the risks mentioned in the current literature far exceed the opportunities.

Summary of Risks and Opportunities
In the context of CAW, excessive demands due to information overload and overwork, monotonous work, restrictions on autonomy, and a lack of privacy can lead to increased tensions between employees or even result in bullying. The same holds true for increased stress due to the perception of surveillance, concerns, and increased

competition at the organisational level due to digital innovations. The non-transparent handling of CAW by supervisors and viewing employees primarily as data can also be problematic for relationships. How relationships in the workplace change in the context of CAW and human digital twins thus also depends significantly on other manifestations of the transformation of work. The shift of a large part of face-to-face communication to digital communication and digital networks, an increased number of employees working from home, and the growing automation and fragmentation of work tasks can lead to increased isolation and weak social relationships. Much is likely to be lost in the use of digital communication that could be achieved with personal contact in terms of social integration. Digital communication could thereby undermine the function of the workplace as a place of social integration. However, the reduction of informal contact through increased work-related digital communication could also undermine the efficiency of work processes, as the personal content of communication that acts as a 'lubricant' might be lost, i.e., trust and familiarity could be more difficult to establish. This risk is likely to apply in particular to people who do not yet know each other personally. The consequences of the occurrence of various risks to relationships could be the absence of a sense of belonging and a decline in intrinsic motivation. Possible downstream effects of low retention could be frequent employee turnover, with costs for employers and a loss of internal knowledge. All the previously mentioned risks could be accompanied by a general decline in productivity in the company.

However, there is also evidence that CAW in general may protect against bullying at the workplace (Melzer & Diewald, 2020). The chances for contact and collaboration through digital communication and digital networks can also increase an employee's sense of belonging. For example, location-independent work contacts can be maintained. Digital networks can also help individuals to feel solidarity with each other and to organise resistance to employers. However, the fact that algorithmic work control might lead to increased solidarity because of the poor working conditions that often accompany it should be viewed very critically. For the employer, a revolt against management resulting from solidarity could lead to a drop in productivity.

Possibilities for Coping with Risks

The risks and opportunities highlight the importance of what functions CAW fulfils, how CAW is implemented, and how employees use technological systems and perceive CAW. The primarily considered design aspects, such as ergonomics and user experiences, are mainly related to individuals. According to our findings, the design of digital technologies and human digital twins for the work context should take into account that the technologies will be used in complex socio-technical systems. These socio-technical systems can affect the social integration of employees and thus workplace relationships. These social relationships eventually affect the individual again. The implementation of CAW depends on how certain functions are handled by employers and supervisors. Ultimately, it is also crucial for how employees deal with and perceive the implemented functions. Based on the literature review and our findings, the following practices that may protect or promote positive workplace relationships can be derived.

When designing and implementing technological systems for information and communication, it may make sense to provide opportunities for self-determined communication and information gathering, for example, via digital networks, to enable more contacts between employees. However, a large part or even all of the communication and information gathering should not be planned from the outset via such functions. In this way, employees can decide to a certain extent how much they meet their need to belong via CAW or personal contact. An example that is already used to solve the problem of the lack of informal contact opportunities in videoconferencing is the addition of tools such as wonder.me, through which employees can talk informally and without being monitored by supervisors. The negative effects that information overload and acceleration can have in connection with CAW point to a need for action that is related to the individual appropriation of CAW by employees. This is where, for example, training for employees on topics such as self-management in the use of digital technologies could come in. Furthermore, comprehensive training on the use of the design aspects of the communication and information functions of digital technologies could lead to employees being able to use these functions more efficiently for their own purposes. These design aspects could possibly be adapted so that employees are able to, for instance, control and filter the flood of information on their own.

To avoid negative consequences for relationships due to the indirect influences of CAW, workplace characteristics such as routine work, autonomy, control, transparency, privacy, and datafication should be taken into account when designing digital technologies for CAW. Functions can be made transparent, so that employees and employers are able to understand how they work. Employee participation regarding the ways in which CAW is implemented might be beneficial for employees. Monitoring functions should be used by supervisors in a way that ensures that they do not interfere with privacy. When implementing CAW, those digital technologies that actively support employee learning can be focused on. Automatically generated feedback could be used primarily or partly for the self-monitoring of employees so that they can improve on their own without being evaluated by supervisors. Moreover, automatically generated feedback should not be used by supervisors as the only source of performance appraisal. A fragmentation of tasks in the context of CAW can mean more routine work for employees. A variety of digital technologies to help with the introduction to other tasks could be used to provide an easy-to-implement regular variety of work tasks. For employees to perform CAW in a primarily self-directed way, they need to have specific knowledge about CAW. This is where extensive training for all groups of employees could again be useful (see Chapter "Which Types of Workers Are Adversely Affected by Digital Transformation? Insights from the Task-Based Approach" which describes the need for further training regarding digital technologies). If employers and supervisors want to prevent resistance against bad working conditions, they should consider implementing CAW in a way that supports employees and fosters social integration.

Conclusions and Directions for Further Research

The potential risks and opportunities identified in this contribution are mostly only rudimentarily explored so far. Most of the literature is based on individual case studies or theoretical assumptions, and it often deals with workplace relationships only as a secondary topic. Future research should address the major research gaps regarding CAW and the associated consequences for social integration in the workplace. Not only the design and implementation of CAW, but also the experience dimension, i.e., how employees perceive and evaluate the design and implementation of CAW, should be considered. We suggest a differentiated consideration of the individual perceptions of the different functions of digital technologies used for CAW and the way they are implemented, focusing on the consequences for workplace relationships. Both risks and opportunities should be taken into account. It seems necessary to examine the immense risks in order to be able to counteract a threat to workplace relationships. The opportunities for CAW to improve workplace relationships have hardly been investigated at all so far, which could explain the imbalance in favour of risks that we found in our review. Therefore, a focus on opportunities that may still need to be identified could bring many new insights. Future (empirical) research should also consider that workplace relationships can be shaped at both the individual and organisational levels in the context of CAW. Distinguishing between relationships with supervisors, relationships with colleagues in intra-departmental teams, relationships with colleagues in inter-departmental teams, cross-company relationships, and de-commissioned relationships among gig and crowd workers could also be useful in generating a more detailed understanding of CAW.

Acknowledgements Sarah Brunsmeier and Martin Diewald are members of the research programme 'Design of Flexible Work Environments—Human-Centric Use of Cyber-Physical Systems in Industry 4.0', which is supported by the North Rhine-Westphalian funding scheme 'Forschungskolleg'.

References

Aiello, J. R., & Kolb, K. J. (1995). Electronic performance monitoring and social context: Impact on productivity and stress. *The Journal of Applied Psychology, 80*(3), 339–353. https://doi.org/10.1037/0021-9010.80.3.339

Apt, W., Schubert, M., & Wischmann, S. (2018). Digitale Assistenzsysteme–Perspektiven und Herausforderungen für den Einsatz in Industrie und Dienstleistungen.

Baillien, E., & de Witte, H. (2009). Why is organizational change related to workplace bullying? Role conflict and job insecurity as mediators. *Economic and Industrial Democracy, 30*(3), 348–371. https://doi.org/10.1177/0143831X09336557

Ball, K. (2010). Workplace surveillance: An overview. *Labor History, 51*(1), 87–106. https://doi.org/10.1080/00236561003654776

Ball, K. (2021). *Electronic monitoring and surveillance in the workplace*. Luxembourg: Publication Office of the European Union.

Bannister, F., & Connolly, R. (2011). The trouble with transparency: A critical review of openness in e-government. *Policy & Internet, 3*(1), 158–187. https://doi.org/10.2202/1944-2866.1076

Berisha-Gawlowski, A., Caruso, C., & Harteis, C. (2021). The concept of a digital twin and its potential for learning organizations. In D. Ifenthale, S. Hofhues, M. Egloffstein & C. Helbig (Eds.), *Digital transformation of learning organizations* (pp. 95–114). Berlin, Springer Open. https://doi.org/10.1007/978-3-030-55878-9_6

Boes, A. (2016). Produktivkraftsprung Informationsraum: Geschäftsmodelle, Wertschöpfung und Innovation neu denken. Disruptive Innovation. Digitalisierung und der Umbruch in der Wirtschaft. In Dokumentation des 13. Innovationsforum der Daimler und Benz Stiftung, Ladenburg.

Boes A, Kämpf T, Ziegler A (2020) Soziologie des Digitalen - Digitale Soziologie? In: Maasen S, Passoth JH (eds) Soziologie des digitalen - Digitale soziologie? Nomos, Baden-Baden, Soziale Welt Sonderband. https://doi.org/10.5771/9783845295008

Bonde, J. P., Gullander, M., Hansen, Å. M., Grynderup, M., Persson, R., Hogh, A., Willert, M. V., Kaerlev, L., Rugulies, R., & Kolstad, H. A. (2016). Health correlates of workplace bullying: A 3-wave prospective follow-up study. *Scandinavian Journal of Work, Environment & Health, 42*(1), 17–25. https://doi.org/10.5271/sjweh.3539

Boswell, W. R., & Olson-Buchanan, J. B. (2007). The use of communication technologies after hours: The role of work attitudes and work-life conflict. *Journal of Management, 33*(4), 592–610. https://doi.org/10.1177/0149206307302552

Castanheira, F., & Chambel, M. J. (2010). Reducing burnout in call centers through HR practices. *Human Resource Management, 49*(6), 1047–1065. https://doi.org/10.1002/hrm.20393

Ciocchetti, C. A. (2011). The eavesdropping employer: A twenty-first century framework for employee monitoring. *American Business Law Journal, 48*(2), 285–369. https://doi.org/10.1111/j.1744-1714.2011.01116.x

Cohen, S. G., & Bailey, D. E. (1997). What makes teams work: Group effectiveness research from the shop floor to the executive suite. *Journal of Management, 23*(3), 239–290. https://doi.org/10.1177/014920639702300303

Colbert, A. E., Bono, J. E., & Purvanova, R. K. (2016). Flourishing via workplace relationships: Moving beyond instrumental support. *Academy of Management Journal, 59*(4), 1199–1223. https://doi.org/10.5465/amj.2014.0506

Collins, A. M., Hislop, D., & Cartwright, S. (2016). Social support in the workplace between teleworkers, office-based colleagues and supervisors. *New Technology, Work and Employment, 31*(2), 161–175. https://doi.org/10.1111/ntwe.12065

Cooper, C. D., & Kurland, N. B. (2002). Telecommuting, professional isolation, and employee development in public and private organizations. *Journal of Organizational Behavior, 23*(4), 511–532. https://doi.org/10.1002/job.145

Crowley, M. (2014). Class, control, and relational indignity. *American Behavioral Scientist, 58*(3), 416–434. https://doi.org/10.1177/0002764213503335

Dambrin, C. (2004). How does telework influence the manager-employee relationship? *International Journal of Human Resources Development and Management, 4*(4), 358. https://doi.org/10.1504/IJHRDM.2004.005044

Diewald, M., & Nebe, K. (2020). Familie und Beruf: Vereinbarkeit durch Homeoffice? *Soziologische und rechtwissenschaftliche Perspektiven. Sozialer Fortschritt, 69*(8–9), 595–610. https://doi.org/10.3790/sfo.69.8-9.595

Einarsen, S., & Mikkelsen, E. G. (2003). Individual effects of exposure to bullying at work. In: S. Einarsen (Ed.), *Bullying and emotional abuse in the workplace*, Taylor & Francis, London and New York (pp. 127–144). https://doi.org/10.1201/9780203164662-12

Einarsen, S., & Nielsen, M. B. (2015). Workplace bullying as an antecedent of mental health problems: A five-year prospective and representative study. *International Archives of Occupational and Environmental Health, 88*(2), 131–142. https://doi.org/10.1007/s00420-014-0944-7

Einarsen, S., Raknes, B. I., & Matthiesen, S. B. (1994). Bullying and harassment at work and their relationships to work environment quality: An exploratory study. *European Work and Organizational Psychologist, 4*(4), 381–401. https://doi.org/10.1080/13594329408410497

Franzen, M. (2021). Bdsg §26 rn. In R. Müller-Glöge, U. Preis, & I. Schmidt (Eds.), *Erfurter Kommentar zum Arbeitsrecht, Beck'sche Kurz-Kommentare* (pp. 27–34). München: C.H. Beck.

Funk, M., Backhaus, N., Terhoeven, J., & Wischniewski, S. (2019). Menschzentrierte Gestaltung digitaler Arbeitsassistenz: Herausforderungen hinsichtlich Überwachung und Datenschutz kontextsensitiver Systeme.

Gal, U., Jensen, T. B., & Stein, M. K. (2020). Breaking the vicious cycle of algorithmic management: A virtue ethics approach to people analytics. *Information and Organization, 30*(2). https://doi.org/10.1016/j.infoandorg.2020.100301

Goll, M. (2008). Professionalisierungs- und Inszenierungsstrategien in der beruflichen Netzkommunikation. In Weltweite Welten, VS Verlag für Sozialwissenschaften (pp. 223–246). https://doi.org/10.1007/978-3-531-91033-8_10

Gräßler, I., & Poehler, A. (2017). Integration of a digital twin as human representation in a scheduling procedure of a cyber-physical production system. In *2017 IEEE International Conference on Industrial Engineering and Engineering Management (IEEM)* (pp. 289–293). IEEE. https://doi.org/10.1109/ieem.2017.8289898

Gräßler, I., & Pöhler, A. (2020). Produktentstehung im Zeitalter von Industrie 4.0. In Handbuch Gestaltung digitaler und vernetzter Arbeitswelten (pp. 383–403). Berlin, Heidelberg: Springer. https://doi.org/10.1007/978-3-662-52979-9_23

Gräßler, I., Roesmann, D., & Pottebaum, J. (2020). Traceable learning effects by use of digital adaptive assistance in production. *Procedia Manufacturing, 45*, 479–484. https://doi.org/10.1016/j.promfg.2020.04.058

Gräßler, I., Roesmann, D., Cappello, C., & Steffen, E. (2021). Skill-based worker assignment in a manual assembly line. *Procedia CIRP, 100*, 433–438. https://doi.org/10.1016/j.procir.2021.05.100

Green, F. (2011). Unpacking the misery multiplier: How employability modifies the impacts of unemployment and job insecurity on life satisfaction and mental health. *Journal of Health Economics, 30*(2), 265–276. https://doi.org/10.1016/j.jhealeco.2010.12.005

Hansen, M., & Thiel, C. (2012). Cyber-Physical Systems und Privatsphärenschutz. *Datenschutz und Datensicherheit - DuD, 36*(1), 26–30. https://doi.org/10.1007/s11623-012-0007-8

Heath, C., Knez, M., & Camerer, C. (1993). The strategic management of the entitlement process in the employment relationship. *Strategic Management Journal, 14*(S2), 75–93. https://doi.org/10.1002/smj.4250141008

Hirsch-Kreinsen, H., & ten Hompel, M. (2017). Digitalisierung industrieller arbeit: Entwicklungsperspektiven und gestaltungsansätze. In: B. Vogel-Heuser, T. Bauernhansl & M. ten Hompel (eds) *Handbuch Industrie 4.0, Springer Reference Technik* (pp. 357–376). Berlin, Heidelberg: Springer Vieweg.

Hodson, R. (1996). Dignity in the workplace under participative management: Alienation and freedom revisited. *American Sociological Review, 61*(5), 719–738. https://doi.org/10.2307/2096450

Kelly, E. L., & Moen, P. (2020). Overload: How good jobs went bad and what we can do about it. *Princeton University Press, Princeton and Oxford,*. https://doi.org/10.1515/9780691200033

Kensbock, J. M., & Stöckmann, C. (2021). "Big brother is watching you": Surveillance via technology undermines employees' learning and voice behavior during digital transformation. *Journal of Business Economics, 91*(4), 565–594. https://doi.org/10.1007/s11573-020-01012-x

Khanchel, H. (2020). New means of workplace surveillance model: From the gaze of the supervisor to the digitalization of employee performance. *Business and Management Research, 8*(4), 54–64. https://doi.org/10.5430/bmr.v8n4p54

Knights, D., & McCabe, D. (2003). Governing through teamwork: Reconstituting subjectivity in a call centre*. *Journal of Management Studies, 40*(7), 1587–1619. https://doi.org/10.1111/1467-6486.00393

Kort, M. (2018). Neuer Beschäftigtendatenschutz und Industrie 4.0. RdA - Recht der Arbeit *71*(1), 24–33. https://beck-online.beck.de/?typ=reference&y=300&b=2018&n=1&s=24&z=RDA

Kunze, F., Hampel, K., & Zimmermann, S. (2020). Homeoffice in der Corona-Krise: Eine nachhaltige Transformation der Arbeitswelt?

Lange, S., Burr, H., Conway, P. M., & Rose, U. (2019). Workplace bullying among employees in Germany: Prevalence estimates and the role of the perpetrator. *International Archives of Occupational and Environmental Health, 92*(2), 237–247. https://doi.org/10.1007/s00420-018-1366-8

Melzer, S. M., & Diewald, M. (2020). How individual involvement with digitalized work and digitalization at the workplace level impacts supervisory and coworker bullying in German workplaces. *Social Sciences, 9*(9). https://doi.org/10.3390/socsci9090156

Michel, A., & Wöhrmann, A. M. (2018). Räumliche und zeitliche Entgrenzung der Arbeit: Chancen, Risiken und Beratungsansätze. *PiD - Psychotherapie im Dialog, 19*(03), 75–79. https://doi.org/10.1055/a-0556-2465

Mikkelsen, E. G., & Einarsen, S. (2002). Relationships between exposure to bullying at work and psychological and psychosomatic health complaints: The role of state negative affectivity and generalized self-efficacy. *Scandinavian Journal of Psychology, 43*(5), 397–405. https://doi.org/10.1111/1467-9450.00307

Moore, P. V. (2018). *The threat of physical and psychosocial violence and harassment in digitalized work*. International Labour Office Geneva.

Moussa, M. (2015). Monitoring employee behavior through the use of technology and issues of employee privacy in America. *SAGE Open, 5*(2). https://doi.org/10.1177/2158244015580168

Niehaus, J. (2017). Mobile Assistenzsysteme für Industrie 4.0: Gestaltungsoptionen zwischen Autonomie und Kontrolle. Forschungsinstitut für gesellschaftliche Weiterentwicklung eV 4.

Nöhammer, E., & Stichlberger, S. (2019). Digitalization, innovative work behavior and extended availability. *Journal of Business Economics, 89*(8–9), 1191–1214. https://doi.org/10.1007/s11573-019-00953-2

Rexroth, M,, Sonntag, K., Goecke, T., Klöpfer, A, & Mensmann, M. (2014). Wirkung von anforderungen und ressourcen auf die zufriedenheit mit der work-life-balance. In: K. Asanger (Eds.), *Arbeit und Privatleben harmonisieren: Life Balance Forschung und Unternehmenskultur: das WLB-Projekt* (pp. 85–128).

Salin, D., & Hoel, H. (2003). Organisational antecedents of workplace bullying. In: S. Einarsen (Ed.), *Bullying and emotional abuse in the workplace*, Taylor & Francis, London and New York (pp. 203–218). https://doi.org/10.1201/9780203164662-17

Schaupp, S. (2022). Algorithmic integration and precarious (dis)obedience: On the co-constitution of migration regime and workplace regime in digitalised manufacturing and logistics. *Work, Employment and Society, 36*(2), 310–327. https://doi.org/10.1177/09500170211031458

Staab, P., & Geschke, S. C. (2020). *Ratings als arbeitspolitisches Konfliktfeld: Das Beispiel Zalando*. Düsseldorf: Hans-Böckler-Stiftung.

Trier, M., & Bobrik, A. (2009). Social search: Exploring and searching social architectures in digital networks. *IEEE Internet Computing, 13*(2), 51–59. https://doi.org/10.1109/MIC.2009.44

Trier, M., & Richter, A. (2015). The deep structure of organizational online networking—An actor-oriented case study. *Information Systems Journal, 25*(5), 465–488. https://doi.org/10.1111/isj.12047

Vallas, S. P. (2006). Empowerment redux: Structure, agency, and the remaking of managerial authority. *American Journal of Sociology, 111*(6), 1677–1717. https://doi.org/10.1086/499909

de Witte, H. (1999). Job insecurity and psychological well-being: Review of the literature and exploration of some unresolved issues. *European Journal of Work and Organizational Psychology, 8*(2), 155–177. https://doi.org/10.1080/135943299398302

Wood, A. J. (2020). *Despotism on demand: How power operates in the flexible workplace*. Ithaca: ILR Press, an imprint of Cornell University Press.

Yerby, J. (2013). Legal and ethical issues of employee monitoring. *Online Journal of Applied Knowledge Management (OJAKM), 1*(2), 44–55.

Zuboff, S. (2019). *The age of surveillance capitalism: The fight for the future at the new frontier of power*. London: Profile Books.

Which Types of Workers Are Adversely Affected by Digital Transformation? Insights from the Task-Based Approach

Talea Hellweg and Martin Schneider

Abstract How digital technologies will affect skills and training needs is one of the conundrums of the evolving Industry 4.0. Important answers have been provided over the past 20 years by a new perspective: the task-based approach within labour economics. This approach interprets a job as a set of tasks, and it exploits employee survey data to measure and compare jobs in terms of skill needs. This chapter presents insights gained from the extant literature within this approach, including our own research, and identifies groups of employees whose job prospects and training opportunities are especially vulnerable. In particular, the evidence suggests that digital technologies often replace human employees who perform routine tasks. The remaining jobs will tend to include more sophisticated tasks such as directing and planning. Employees in jobs involving more routine tasks will therefore most likely need considerable retraining and upskilling. For various reasons, employers are unlikely to support this group of employees in terms of training and retraining. At the same time, routine jobs are often highly specific—their sets of tasks differ from those of most available jobs. Therefore, employees with such skill profiles are less likely to easily find alternative jobs. In other words, we identify a triple risk of digitalisation for some workers, namely those who conduct routine tasks, have a very specific skill profile that limits their ability to move to other jobs, and receive little training and retraining. Our approach and findings have important managerial and policy implications. Based on the framework we suggest, employers could identify and support employees with considerable risk by tracking their skills in a competence management system, i.e., by examining the digital twins of employees in terms of skills. If employers are unlikely to offer enough further training and retraining, an appropriate public policy response would be to reorganise initial vocational training—more than in the existing system, young employees need to be equipped with an even more general set of tasks, which would allow them to switch jobs within (and perhaps across) occupations.

T. Hellweg (✉) · M. Schneider
University Paderborn, Warburger Str. 100, 33098 Paderborn, Germany
e-mail: talea.hellweg@uni-paderborn.de

M. Schneider
e-mail: martin.schneider@uni-paderborn.de

231

Keywords Technological change · Job skills · Task-based approch · Human digital twin of employees · Further training

1 Introduction

It is regularly predicted that Industry 4.0 technologies will eliminate millions of jobs and change the skills that workers need to perform the jobs that remain. These fundamental shifts have become common knowledge and raised considerable concerns among employees, employers, and public policy-makers alike. However, these changes will not occur overnight. They have been going on for years, so labour market data can be utilised to address urgent questions about the future of work: Which skills will employees need when they work with new technologies? Which types of jobs are threatened by automation? How likely are groups of workers to find a new job when their current job is at risk due to digitalisation? Which types of workers need which types of training and retraining?

This chapter presents answers to these questions by reviewing key findings (including our own research) of the so-called 'task-based approach' (Autor et al., 2003). Since the beginning of the century, extremely fine-grained data in which employees provide information about what they actually do in their jobs (for example, writing or handling a machine, performing calculations, operating a computer, teaching, or giving directions to other workers) have been available. The task-based approach utilises these data to describe the skill profiles that employees need in order to do their jobs, to examine how these profiles change when new technologies are used, and to determine which occupations or jobs are becoming obsolete (Acemoglu & Autor, 2011; Autor et al., 2003; Eisele & Schneider, 2020; Goos et al., 2014; Spitz-Oener, 2006).

The task-based approach matters in the context of this book in two ways. First, based on the findings in the literature, it is possible to predict the likely changes in job requirements that digitalisation will bring about and to identify groups of workers whose employment prospects and training opportunities are especially at risk. This is crucial information for employees and their representatives, employers, and public policy-makers. It is important for choosing certain jobs, planning further training and retraining for employees, reorganising job profiles within the German dual vocational training system, and curbing the effects of unemployment and pay cuts for disadvantaged groups of employees.

Second, the findings of the task-based approach suggest that the digital twins of employees can be utilised to contribute to a skills and people management system dedicated to long-term employment. A digital twin is a digital replica of a real-world object, and the human digital twin is a digital replica of a person (see Part I). Most larger companies record the competencies of their employees in digital database structures. The systems thus create human digital twins of employees. They have been introduced in order to better match work tasks with employees and to identify employee training needs. In light of the results of the task-based literature, employers can use and refine their competence management systems to identify workers with particular risks in terms of automation or changes in skill profiles. They can then offer targeted training or retraining to prepare their employees for the new work requirements associated with digitalisation. Though the data involved are sensitive and need to be protected appropriately, our chapter illustrates that the digital twins of employees can be applied in ways that benefit both employers and employees.

This chapter reviews and contributes to literature that is based on regression-based secondary analyses of larger questionnaire surveys of employees. The main advantage of this is that fine-grained information on task profiles across a whole range of occupations is available. This information can then be linked to whether workers have been affected by digitalisation or have received training. The most important result of this chapter is an unlucky interaction of three risk factors: conducting so-called routine tasks, having a very specific skill profile that limits movement to other jobs, and receiving little training and retraining. More particularly, employees with routine tasks are especially affected, as they are at high risk of having their work tasks automated. Therefore, they need retraining and upskilling to meet new, more complex work requirements such as directing and planning (Lukowski et al., 2020). Workers with specific skill profiles are additionally affected because their task profiles differ from those on the labour market, making job changes more difficult (Lazear, 2009; Eggenberger et al., 2018). It is consequently harder for them to adapt flexibly and find an alternative job when work tasks or jobs are eliminated by digitalisation. In order to prevent unemployment or a loss of wages due to lower adaptability, these employees are also in particular need of retraining or further training. However, for various reasons, employers at the moment have little incentive to invest in the education and training of these employee groups. To prevent systematic disadvantages for these groups, one possible public policy response could be redesigning initial vocational training. The task-based framework implies that vocational training must equip young workers with the ability to perform a set of general tasks, enabling them to move more easily within and between occupations.

Section 2 introduces how the task-based approach evolved from the study of skill-biased technological change in economics. Based on this research, Sect. 3 derives a matrix of risks that employees with differing skills are exposed to. Section 4 argues that employees with the highest risk are those who are least likely to receive training and retraining by employers. Section 5 outlines the use of human digital twins in an HR context to mitigate the risks of digitalisation. Section 6 summarises the policy implications derived from this research.

2   Changing Work Through Digitalisation: From Skill-Biased Technological Change to the Task-Based Approach

The skill-biased technological change (SBTC) approach was the main perspective in economics that was used to study how technology affects the labour demand (Acemoglu & Autor, 2011). When new technologies such as computers, new information and communication technologies, or robots become cheaply available, they either replace or complement human labour. In the first case, the demand for human labour services decreases as a result of new technologies, and in the latter case, the demand may increase (Acemoglu & Restrepo, 2018; Dauth et al., 2017). The effect that dominates differs between the skill levels required for various jobs (Arntz et al., 2016). Industrial robots, for example, replace human machine operators but require more engineers (Dauth et al., 2017).

The main finding of research using the SBTC approach is an observed shift in employee demand toward more highly skilled workers. It has been argued and demonstrated empirically that it is primarily employees with a lot of training and with formal university and similar degrees who benefit from the use of information and communication technologies (ICT), while the jobs of workers with lower levels of education are more likely to be performed by ICT and other new technologies. As a result, income inequality will tend to rise (Katz & Murphy, 1992). The task-based approach is a continuation of the SBTC perspective. It also examines how technological transformation changes labour demand, but it shifts attention away from formal qualification towards the skill or task profiles that make up a vocation or a particular job. The empirical analysis in the task-based approach builds on progress in questionnaire surveys of employees. Since the beginning of the century, very fine-grained data in which workers report on their jobs and on the technologies they use and are surrounded by have become available. Now, researchers can examine how new technologies affect different types of jobs over time—they may replace them, render them more important, or change their demands in terms of the work tasks involved. Hence, this approach is grounded in the assumption that the fundamental laws of substitution and complementarity are always at work, meaning that we can infer the future of jobs and occupations from what happened in the recent past.

In this context, a 'work task' is a generic type of activity such as teaching, developing and researching, operating a machine, giving instructions, or talking to a customer. It is assumed that each job held by an individual involves a number of such activities to differing degrees. In questionnaire surveys, employees report what they actually do on a daily basis and how important each activity is in terms of its time span. Though it may be crude to infer the importance of an activity from how much time it takes, using this type of data is the closest research can get to describing job content. Building on this information, researchers are able to derive valid descriptions of workers' jobs in the form of profiles of tasks and, by implication, profiles of the skills necessary for conducting these tasks. Researchers can also describe different occupations in similar ways: What a bicycle mechanic, a plant breeder, a loan officer, or a nurse typically does can be summarised by the activities ('tasks') involved and the importance of each task.

Importantly, the task-based approach posits that the nature of the work tasks involved in a job also determines whether or not a particular job or occupation will be eliminated or changed by digitalisation (Acemoglu & Autor, 2011; Autor et al., 2003). By identifying elimination and change in the past, the literature also seeks to identify which groups of employees will be at a disadvantage in the future. This is crucial information if we want to alleviate these disadvantages.

In summary, by focusing on work tasks, the task-based approach has considerably enriched the SBTC perspective, which studied shifts in the demand for workers depending on different levels of formal education and skills (Acemoglu, 2002; Katz & Murphy, 1992; Machin & van Reenen, 1998). The typical question in the SBTC perspective was how unskilled workers compared to workers with a university degree or with a formal vocational degree. The task-based approach allows for more itemised analyses by comparing, for example, patent lawyers with railroad engineers (who both hold university degrees).

The main finding of the SBTC approach, as mentioned above, was a general shift in the labour demand toward more highly skilled workers. The task-based approach uncovered a new pattern—as a result of digitalisation, labour market demand is becoming more polarised (Acemoglu & Autor, 2011; Goos et al., 2009, 2014). 'Polarisation' in this context means that employees at the upper and lower ends of the education spectrum achieve more favourable labour market outcomes than workers in the middle. When their work environment becomes computerised, unskilled or semi-skilled workers and workers with university degrees are often able to retain their wage levels and their jobs. By contrast, employees in the middle of the educational spectrum often lose their jobs due to automation and tend to receive lower wages when their job is retained. In Germany, for example, this often affects workers in the manufacturing sector with a formal degree from the vocational training system, the duale Berufsausbildung (Arntz et al., 2016). Of course, this polarisation does not necessarily apply to every occupational group and in every country. In the administrative and service sectors in Germany, for example, digitalisation has led to a stronger demand for specialists and highly qualified workers, at the expense of simple tasks (Arntz et al., 2016). Nevertheless, the polarisation theory generally reflects observable developments in the labour market.

One of the strengths of the task-based approach is that it allows researchers to identify possible reasons for polarisation. The most important answer the approach has produced is the routine-bias hypothesis (Acemoglu & Restrepo, 2018; Autor et al., 2003; Goos et al., 2014; de Vries et al., 2020). It considers how many tasks within a job are routine as an important criterion for predicting potential workplace automation. Routine tasks are those that are repetitive and can be converted into computer code. The routine-bias hypothesis argues that mainly workers with medium qualifications and medium wages perform routine tasks, so that the automation of routine jobs leads to a polarisation of the labour market. To investigate this empirically, work tasks are classified based on their 'routine degree'. Depending on the study, the

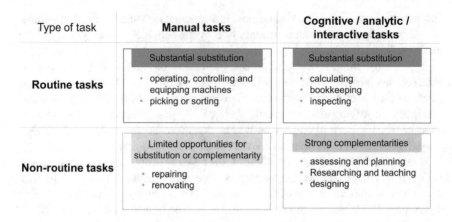

Type of task	Manual tasks	Cognitive / analytic / interactive tasks
Routine tasks	Substantial substitution • operating, controlling and equipping machines • picking or sorting	Substantial substitution • calculating • bookkeeping • inspecting
Non-routine tasks	Limited opportunities for substitution or complementarity • repairing • renovating	Strong complementarities • assessing and planning • Researching and teaching • designing

Fig. 1 Task categories and the risk of substitution. *Note* Figure based on Autor et al. (2003)

classification may differ. A frequently chosen framework has four task categories: routine manual, non-routine manual, routine cognitive/analytic/interactive, and non-routine cognitive/analytic/interactive (Autor et al., 2003; Spitz-Oener, 2006; de Vries et al., 2020). Figure 1 illustrates the different task categories with a few examples and shows the potential for them to be replaced or complemented by new technologies.

Among these tasks, the routine manual tasks are those that are most often eliminated due to digitalisation. Examples of these tasks are operating or controlling and equipping machines. Routine cognitive and analytical tasks such as performing calculations or bookkeeping can often be replaced as well. In contrast, non-routine manual tasks such as repairs or renovations are difficult to replace with technology and will therefore remain. The same is true for non-routine cognitive or analytical tasks such as assessment and planning. These tasks, along with interactive tasks such as teamwork, are becoming more relevant in future jobs (Deming, 2017; Eisele & Schneider, 2020; Spitz-Oener, 2006).

Though the polarisation proposition has become widely accepted, the routine-bias hypothesis, which explains it, is subject to some criticism (Sebastian & Biagi, 2018). In particular, the classification of tasks into routine and non-routine tasks, as well as manula and cognitive tasks, in different papers is partly based on different classification and operationalisation procedures, since the definitions of task types are not uniformly specified (Sebastian & Biagi, 2018). In addition, the assignment of work tasks to routine or non-routine categories can also change due to technological progress (Arntz et al., 2019). Many activities that appeared to be unaffected by automation 20 years ago are now considered to be automatable in the foreseeable future; a well-known example is driving a car (Arntz et al., 2019). Machine learning (ML) methods are reducing the number of tasks that cannot be automated (Arntz et al., 2019; Brynjolfsson & Mitchell, 2017). Long-term reliable forecasts are thus becoming more difficult. Moreover, the distribution of routine occupations in some countries or occupational groups may be at the bottom rather than in the middle of the wage hierarchy, so that occupational upgrading rather than polarisation may be the consequence of the elimination of routine tasks (Haslberger, 2021).

Though these problems should be kept in mind, the task-based approach is of considerable practical importance and offers useful tools for investigating possible changes in work tasks and thus in the labour demand. In what follows, previous work on the task-based approach, including our own, is used to identify groups of employees who are particularly vulnerable in terms of future wages and the labour demand.

3 Digitalisation Effects: Linking the Routine-Bias Hypothesis to Insights on Worker Mobility

In order to identify how different groups of employees are affected by digitalisation, two fundamental aspects are outlined below, namely, the number of routine tasks, which affects the likelihood that work tasks will be automated (Sect. 3.1), and the specificity of the work tasks, which affects workers' mobility (Sect. 3.2). Both aspects represent possible risk factors for employees due to digitalisation. A combination of these aspects is particularly disadvantageous, as discussed in Sect. 3.3.

3.1 Routine Tasks and Automation

The extent to which the automation of routine tasks will lead to the loss of jobs in the future is much debated in the literature. Frey and Osborne caused quite a stir with their study on job loss and the resulting prediction that about 47% of total US jobs are at risk (Frey & Osborne, 2017). Further studies have been conducted to investigate whether these drastic results hold true, including for other countries. As it turns out, the findings vary considerably, especially depending on whether occupations or jobs are considered (Arntz et al., 2019). Frey and Osborne (2017) used occupation-level data, and studies using job-level data (a more disaggregated/detailed level) produced less dramatic results. Thus, for the U.S., Nedelkoska (2013) predicted that only 10% of U.S. workers are in the 'high-risk' group, and Arntz et al. (2017) found that 9% of all U.S. workers are performing automatable jobs. For Germany, studies using job-level data also yield significantly lower values compared to the value of 47% predicted by Frey and Osborne (2017). Dengler and Matthes (2018) concluded that 15% of all employees in Germany belong to the 'high-risk group'. The study by Arntz et al. (2016) provides a similar estimate of 12% (for more detailed information, see Arntz et al. (2019)).

The different study results illustrate how difficult it is to make a correct risk prediction. Arntz et al. (2019) additionally argued that most of the results are frequently upward-biased for two reasons. First, they ignore the fact that workers in seemingly automatable occupations already perform hard-to-automate tasks. Second, the numbers also refer to what could be automated from a technical perspective, not which levels of automation are profitable. Hence, the percentages should not be equated with job losses. Developments to date also show that automation does not lead to the elimination of entire occupations but only to the elimination of individual work tasks (Spitz-Oener, 2006). Additionally, even though the listed studies predict the replacement of human workers by digital technologies and job losses, it can be assumed that digitalisation will also create many new jobs, so that the displacement effects will be partially compensated (Acemoglu & Restrepo, 2018, 2019; Dauth et al., 2017). Nevertheless, work tasks will undoubtedly change in many jobs and occupations. Arntz et al. (2019) therefore concluded that it is not the number but the structure of jobs that will be crucial in the future and that the supply side, namely the education and training system, will have to adapt to the changes in the labour demand.

As mentioned already, the potential for these task changes and the automation of work tasks is not equally distributed across all population groups. Even though a polarisation of the labour market is expected, studies show that low-skilled workers, as opposed to highly skilled workers, are particularly at risk. Gardberg et al. (2020) estimated the average automation probability of low-skilled workers to be almost twice as high as that of university graduates. Therefore, low-income workers are also on average more at risk than high-income workers (Arntz et al., 2016; Nedelkoska, 2013). Nedelkoska and Quintini (2018) additionally showed that the highest automatability is found among jobs held by young employees. For workers aged 30–35 years old, the risk is lowest, and it then increases with age (U-shape). It should be emphasised that low- and medium-income workers will tend to suffer most from the automation of work tasks, while high-income workers may benefit. This systematic effect can contribute to an increase in social inequality and the associated negative effects (Arntz et al., 2016, 2019).

3.2 Specific Task Profiles and Low Mobility

The risk to employees from digitalisation should not be determined solely by the ability to automate their work tasks. Often, digitalisation is accompanied by new tasks and calls for other skills, so it is crucial for employees to be able to adapt to these changes (Bechichi et al., 2019; Nedelkoska, 2013). This adaptability is influenced not only by personal characteristics such as age and job duration, but also by the composition of work tasks and the structure of the labour market. Adaptability is often measured by labour market mobility, i.e., how easy it is for an employee to find a new job if he or she is affected by involuntary layoffs (for example, due to automation).

Mobility is in turn influenced by the work tasks that make up a worker's job or occupation. This idea has been investigated in an interesting stream of research that extends human capital theory in economics. The original theory argued that two types of job-related skills can be distinguished by their marketability: specific and

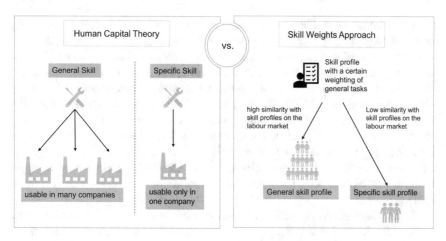

Fig. 2 Graphical comparison of two human capital theories: The left side shows the human capital theory of Becker (1975), and the right side shows the skill weights approach of Lazear (2009)

general 'human capital' (Becker, 1975). General human capital refers to the earning potential of knowledge, skills, and abilities that are of interest to many employers. Being able to work with standard office software is an example of this type of skill. Specific human capital, by contrast, refers to the earning potential of skills that are of interest to one employer only. An example of this type of skill is operating a unique machine. In a recent variation of this theory, it is assumed that each and every skill is generic and therefore general—but that the composition of skills is more or less specific (Lazear, 2009). For example, all journalists need to research, write, and proofread, but a particular job may involve a lot of proofreading and almost no research or writing. This job is then highly specific within the journalist occupation. More generally, individual jobs can be described as more or less specific based on the composition of tasks, i.e., the types of skills involved and the weights that indicate how much each skill is used in the job (Fig. 2).

The idea of different skill weights that influence specificity can help to explain why some workers are more mobile than others (Backes-Gellner & Mure, 2004; Lazear, 2009). If the combination of work tasks is similar to many other combinations on the labour market, it is very general and permits workers who hold this combination to switch easily to other jobs without wage loss. Conversely, when the combination is very unusual or specific, a change of job is more often associated with a longer unemployment period or wage loss (Geel & Backes-Gellner, 2011; Eggenberger et al., 2018).

Specific skill profiles are often connected with higher wages because employees with these skills are difficult to replace, so these workers have an advantage during wage negotiations. Hence, human capital specificity is of value as long as work tasks remain unchanged and the employer is likely to remain in business (Lazear, 2009). However, when jobs become obsolete or work tasks change due to digitalisation, specificity becomes problematic. Workers are more mobile and employable if they have a more general skill profile (Bechichi & Jamet, 2018; Bechichi et al., 2019).

Our own empirical study based on the IAB (Institut für Arbeitsmarkt- und Berufsforschung) panel WeLL dataset ('Further Training as a Part of Lifelong Learning') also assumes that more specific skill profiles lead to fewer opportunities to change jobs or employers; in other words, they lead to lower mobility. Based on this context, we investigate, among other things, which occupational groups are particularly affected by specific skill profiles. The specificity is determined by the distance of the individual task profile from the other profiles of the same occupational group. We find that employees with many routine tasks and a low skill level have, on average, more specific skill profiles and thus fewer opportunities to change jobs (Hellweg, 2021).

The results of Blien et al. (2021) also indicate this same relationship, as they show that workers in routine-intensive occupations experience persistently worse labour market outcomes for the rest of their working lives if their career was affected by an exogenous shock such as displacement during a mass layoff. This can be explained by the lower mobility or employability of workers with routine tasks.

Hence, in order to ensure long-term employment, it is becoming increasingly important to train employees in a variety of skills. Education and training should therefore not be specific; instead, they should be broadly oriented. Employees with very specific skill profiles should therefore take advantage of training offers at an early stage in order to adapt to the requirements of the labour market.

3.3   A Matrix of Risks

The previous subsections have identified routine tasks and skill specificity as risk factors for employees experiencing digitalisation. Figure 3 summarises the two factors using a matrix of worker vulnerability. Each of the four quadrants describes a combination of task routine and skill specificity that predicts labour market outcomes in terms of job and wage loss due to digitalisation.

In quadrants 1 and 4, employees are at a medium level of risk. In both quadrants, workers possess one risk-increasing and one risk-reducing factor. In quadrant 1, employees conduct highly routine tasks, but these tasks are not very specific, so these employees are highly mobile. If routine tasks are eliminated in the course of digitalisation, they can reorient themselves more easily in the labour market.

Quadrant 4 describes the opposite case—the employees perform few routine tasks, but their task profiles are specific, which is why their mobility is low. Accordingly, they are less likely to find another job if they experience job loss, but the probability of job loss due to automation is low.

Employees in quadrant 3 are the most prepared for digitalisation. Their jobs include non-routine tasks and are therefore less likely to be automated in the first place. In the case of job loss, they are highly mobile because of their general skills.

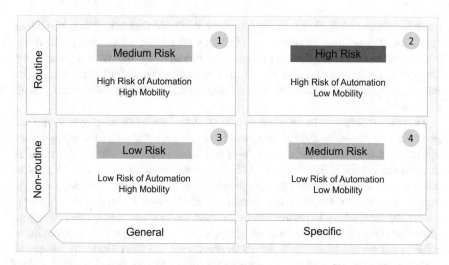

Fig. 3 Job risk properties in times of digital transformation

The situation is reversed for employees in quadrant 2. Their jobs consist of routine tasks that are likely to be automated, and their strongly specific skill profiles render them less mobile than other employees.

This group of high-risk employees deserves the attention of public policy-makers. As the evidence reviewed in Sects. 3.1 and 3.2 shows, mainly low- and middle-skilled and low-income workers have a high risk of being replaced by automation and low mobility (for example, see Bechichi et al. (2019)). Furthermore, the two underlying risk factors are correlated, as our own research based on the WeLL dataset indicates: Employees with many routine tasks often have specific qualification profiles (Hellweg, 2021).

As a result, automation can have a drastic impact on the lower income segment of the labour market.

4 Employees with High Risks Tend to Receive Less Training

Now that we have demonstrated that employees will be affected to varying degrees by digitalisation, this section will discuss which groups of employees actually receive further training to compensate for these risk factors. Training is an important instrument for helping workers transition to new career opportunities when their jobs are at risk of being eliminated by new technologies (Nedelkoska, 2013). Therefore, training can help employees not only to learn how to use concrete new technologies or machines, but also to learn general skills that will be needed in the future (Bechichi et al., 2019). The resistance of employees to the introduction of technologies can also be reduced through further training (Arntz et al., 2016; Ouadahi, 2008; Venkatesh et al., 2002). Moreover, Boothby et al. (2010) showed that the success of the introduction of new technologies, as measured by productivity, increases with training. Due to the importance of training in the process of digital transformation, companies that invest heavily in digitalisation also invest more in training (Arntz et al., 2016; Seyda et al., 2018).

However, study results indicate that it is those groups of employees who are particularly affected by digitalisation that have less access to further training opportunities. Organization for Economic Co-operation and Development (OECD) studies confirm the contradiction that low-skilled workers are most in need of training to adapt to a digitalising workplace but are less likely to engage in firm-based training than other workers (OECD, 2019). Across all OECD countries, highly skilled workers are the most likely to participate in training: On average, nearly 75% of highly skilled workers participate in training, compared with an average of 55% of medium-skilled workers and 40% of low-skilled workers (OECD, 2019). Germany has one of the largest participation gaps between highly skilled and low-skilled workers among OECD countries (OECD, 2021). The results of Mohr et al. (2016) and Bilger et al. (2017) also indicate that low-skilled workers receive less training than highly skilled workers. Mohr et al. (2016) further found that within the group of low-skilled workers, there are differences in training retention, which can be partly explained by the performance of routine tasks.

A study based on data from the National Educational Panel confirms that employees with a high share of routine activities participate in continuing education significantly less often than employees with less routine activities (Heß et al., 2019). Similarly, Lukowski et al. (2020) showed that workers who perform complex tasks more frequently receive more training, regardless of the general skill requirements of the job.

Employees with low mobility can also significantly benefit from training by increasing their adaptability (Bechichi et al., 2019). In the case of constant labour market conditions, employers have high incentives to invest in the human capital of employees with specific skill profiles because, first, they are rarely found in the labour market and, second, they are more likely to remain with the company in the long run due to their low mobility (Eggenberger et al., 2018; Geel et al., 2011; Lazear, 2009). In the case of changing work tasks as a result of digitalisation, however, this positive negotiating position is likely to break down. If the skill profiles are very specific, the required retraining in the context of digitisation may be more cost-intensive, as the old skill profile is more likely to differ from the new requirements. Employers must therefore weigh the costs and benefits. In addition, high adaptability is seen as a competitive advantage in times of digital transformation, and low adaptability reduces the likelihood that employees will be able to stay with the company in the long term. These factors mean that, from a theoretical perspective, employers have less incentive to invest in training for employees with specific skill profiles, even though these employees' need for training is greater, because training costs may exceed the benefits of the long-term employment of these employees.

Our own research with the WeLL data (see above) supports this proposition. Employees with specific work tasks receive significantly less training than those with general skill profiles if their work is affected by the introduction of ICT or new technologies or machines (Hellweg, 2021). This can be explained by the lower bargaining power of workers with specific skill profiles when their work tasks change.

The fact that employees with high risks also receive less training results from an unfortunate accumulation of risk factors. As a result, the structure of training in its current form contributes to a widening of the inequality between highly skilled and low-skilled workers rather than reducing it (Mohr et al., 2016).

It should be noted that the empirical research presented in this paper refers to formal training. Informal training or learning (for example, workplace-based training) can also play a decisive role in helping employees adapt to new work requirements in the context of digitalisation (Dehnbostel, 2018). Informal learning is implicit, unintentional, opportunistic, unstructured, and lacks a teacher (Eraut, 2004). Because of these characteristics, it is difficult to measure, so there is a lack of available data and research. For this reason, it is unclear whether or not the high-risk workers receive more informal training to compensate to some extent for their lack of formal training in response to digitalisation. This issue has been left out in this chapter.

5 Reducing Risks Due to Digitalisation by Using Human Digital Twins in Personnel Management

New technical possibilities also offer companies, especially human resources (HR) departments, opportunities to reduce the risks experienced by employees due to digitalisation outlined in the previous sections. Two technical tools that could help reduce these risks, namely competency management systems and online career platforms, are discussed below. Both tools are based on the creation of information profiles for employees and employers—in other words, on the creation of human digital twins.

Although the previous sections have shown that companies are less likely to train high-risk employees, they are often fundamentally interested in long-term employment, for example, due to transaction costs or corporate culture, and thus they are interested in maintaining the employability of employees. To achieve this, competence management systems can be a helpful tool. They are already used to plan training needs at an early stage and to improve internal job matching (Lindgren et al., 2004; North et al., 2018). They consist of digital representations of employee competencies, as well as job descriptions. The employee profiles can be matched with future requirements to identify training needs. They also enable a comparison with skill profiles in the external and internal labour markets (Lindgren et al., 2004; Miranda et al., 2017). In this way, job changes within the company can be targeted and the internal matching of employees and work tasks can be improved.

Although most companies operate competence management systems, it is unlikely that they utilise the important distinctions between routine/non-routine tasks and general/specific tasks. However, by leveraging what we know about employees that are at risk because of digitalisation, HR managers could develop competence management systems into a tool that could curb the negative effects of digitalisation, not only in terms of skills shortages but also in terms of planning the transition to new technologies and training for particularly vulnerable employees.

For such systems to work, comprehensive data on employee skills and characteristics, as well as on production processes, must be collected, stored, and evaluated. Information can be provided by the employee and their superiors, or it can be generated automatically. To increase the acceptance of data collection, it is helpful if the process is transparent and employees have control over the stored data concerning them (Lindgren et al., 2004). This is especially important if profile entries are not created by the employees themselves (Lindgren et al., 2004). Transparency for others, meaning that profiles can be viewed by colleagues and superiors, can increase the benefits of profiles through the transfer of expertise and sharing of knowledge, but it can also reduce the acceptance of data collection due to a loss of privacy (Lindgren et al., 2004).

Data collection and storage represent a considerable practical challenge and also involve the risk of data misuse if sufficient data protection cannot be guaranteed. The practical implementation of such systems is therefore associated with considerable costs. Companies must determine whether the use of such systems is actually profitable for their application context.

A second important aspect in the context of digitalisation is that matching and mobility in the labour market are becoming increasingly important due to rapidly changing requirements (Arntz et al., 2019). Online career platforms such as LinkedIn, Xing, and Glassdoor are becoming increasingly important in the matching process (Madia, 2011). Employers can present themselves on these platforms and indicate which jobs they are offering and which employees they are looking for. Employees, on the other hand, can present their skills, experience, and interests (Chiang & Suen, 2015; Dutta, 2010). This creates a form of digital twin for both the employee and the employer. A high level of transparency and regional independence on the platform should make it much easier for the two parties of interest to find each other than it would be via more traditional recruiting channels. Simplified matching is especially an advantage for employees who could potentially experience job loss due to digitalisation, as it increases the likelihood of finding an alternative job with similar work requirements. This means that it can be especially advantageous for high-risk workers to be familiar with and participate in such online platforms.

The functions of such professional networks go beyond the classic fact sheet. They serve as a network for business contacts and as a media channel for companies. However, it must be considered that sensitive personal data are stored and published (Dutta, 2010). Even if the data are provided on the platform by the users themselves, these data require special protection. If this protection is not adequately guaranteed, the risks can outweigh the benefits of the digital twins that are created.

In summary, the creation of digital twins in the area of human resources management certainly opens up opportunities to reduce the negative effects of digitalisation by improving job matching and identifying training needs. However, data collection also involves the risk of data misuse if data protection is not adequately guaranteed. Additionally, possible erroneous data evaluation or data-driven discrimination in the application of artificial intelligence (AI) must be critically considered (Ore & Sposato, 2021).

6 Conclusions

The information presented in this chapter shows that some groups of employees not only have a higher risk of being affected by automation due to different characteristics such as routine tasks or low mobility, but also suffer from additional disadvantages because they receive less training. Such structural disadvantages may not only negatively affect employees but also weaken the overall economic system (Feldmann, 2013). A lack of adaptability reduces the speed of digital change and can lead to competitive disadvantages compared with countries with more flexible labour market structures (Lamo et al., 2011). Moreover, an increase in income inequality and unemployment can reduce economic performance (Blien et al., 2021), and these effects are accompanied by an increase in social spending.

Therefore, the correlation of risks we have described should prompt policy-makers to support exactly these groups of employees in order to prevent macroeconomic consequences such as unemployment and productivity decline. This support can include appropriate training programs that compensate for the lack of investment incentives for employers for these occupational groups (Bechichi & Jamet, 2018; Bechichi et al., 2019). Dauth (2020) showed that training subsidies for low-skilled employees significantly increase the cumulative employment duration and earnings in the short- and middle-term for participants. A vocational training landscape that focuses on flexible and holistic task profiles can also contribute to early risk reduction (Eggenberger et al., 2018). In this way, new employees can be trained from the outset in a way that is geared towards rapidly changing requirements; such a training landscape can act as a preventive measure against digitalisation risks.

Not only policy-makers but also employers may be interested in the information on potential risk factors identified in this chapter. By knowing about risk groups, they can initiate, at an early stage, education and training programs adapted to the needs of their high-risk employees. New technologies that use the human digital twins of employees and employers can also help mitigate risks by targeting training needs and improving job matching. However, their application also poses risks, such as data misuse or erroneous and discriminatory analyses.

Finally, the results on risk factors presented here can benefit employees. By knowing their own risk factors, they can seek out retraining and continue their education on their own, independent of their employer's support. For this reason, it is important to transfer knowledge about risk factors to the employees themselves. Finally, employees, including those with routine tasks and specific skill profiles, should consider engaging with online platforms that are about to become more important in the matching process that takes place in the labour market.

References

Acemoglu, D. (2002). Technical change, inequality, and the labor market. *Journal of Economic Literature, 40*(1), 7–72. https://doi.org/10.1257/0022051026976

Acemoglu, D., & Autor, D. (2011). Skills, tasks and technologies: Implications for employment and earnings. *Handbook of Labor Economics, 4*, 1043–1171. https://doi.org/10.1016/S0169-7218(11)02410-5

Acemoglu, D., & Restrepo, P. (2018). The race between man and machine: Implications of technology for growth, factor shares, and employment. *American Economic Review, 108*(6), 1488–1542. https://doi.org/10.1257/aer.20160696

Acemoglu, D., & Restrepo, P. (2019). Automation and new tasks: How technology displaces and reinstates labor. *Journal of Economic Perspectives, 33*(2), 3–30. https://doi.org/10.1257/jep.33.2.3

Arntz, M., Gregory, T., Jansen, S., & Zierahn, U. (2016). *Tätigkeitswandel und Weiterbildungsbedarf in der digitalen Transformation*. Mannheim: ZEW-Gutachten und Forschungsberichte.

Arntz, M., Gregory, T., & Zierahn, U. (2017). Revisiting the risk of automation. *Economics Letters, 159*, 157–160. https://doi.org/10.1016/j.econlet.2017.07.001

Arntz, M., Gregory, T., & Zierahn, U. (2019). Digitization and the future of work: Macroeconomic consequences. In K. F. Zimmermann (Ed.), *Handbook of Labor, Human Resources and Population Economics, Springer eBook Collection* (pp. 1–29). Cham: Springer. https://doi.org/10.1007/978-3-319-57365-6_11-1

Autor, D. H., Levy, F., & Murnane, R. J. (2003). The skill content of recent technological change: An empirical exploration. *The Quarterly Journal of Economics, 118*(4), 1279–1333. https://doi.org/10.1162/003355303322552801

Backes-Gellner, U., & Mure, J. (2004). The skill-weights approach on firm specific human capital: Empirical results for Germany. *SSRN Electronic Journal*. https://doi.org/10.2139/ssrn.710441

Bechichi, N., & Jamet, S. (2018). Moving between jobs: An analysis of occupation distances and skill needs. *OECD Science, Technology and Industry Policy Papers, 52*. https://doi.org/10.1787/d35017ee-en

Bechichi, N., Jamet, S., Kenedi, G., Grundke, R., & Squicciarini, M. (2019). Occupational mobility, skills and training needs. *OECD Science, Technology and Industry Policy Papers, 70*. https://doi.org/10.1787/23074957

Becker, G. S. (1975). *Human capital: A theoretical and empirical analysis, with special reference to education, Human behavior and social institutions* (2nd ed., Vol. 5). New York, NY: Columbia University Press.

Bilger, F., Behringer, F., Kuper, H., Schrader, J., et al. (2017). Weiterbildungsverhalten in Deutschland 2016: Ergebnisse des Adult Education Survey (AES). W. Bertelsmann Verlag.

Blien, U., Dauth, W., & Roth, D. H. (2021). Occupational routine intensity and the costs of job loss: Evidence from mass layoffs. *Labour Economics, 68*(101953). https://doi.org/10.1016/j.labeco.2020.101953

Boothby, D., Dufour, A., & Tang, J. (2010). Technology adoption, training and productivity performance. *Research Policy, 39*(5), 650–661. https://doi.org/10.1016/j.respol.2010.02.011

Brynjolfsson, E., & Mitchell, T. (2017). What can machine learning do? Workforce implications. *Science (New York, NY), 358*(6370), 1530–1534. https://doi.org/10.1126/science.aap8062

Chiang, J. K. H., & Suen, H. Y. (2015). Self-presentation and hiring recommendations in online communities: Lessons from LinkedIn. *Computers in Human Behavior, 48*, 516–524. https://doi.org/10.1016/j.chb.2015.02.017

Dauth, C. (2020). Regional discontinuities and the effectiveness of further training subsidies for low-skilled employees. *ILR Review, 73*(5), 1147–1184. https://doi.org/10.1177/0019793919885109

Dauth, W., Findeisen, S., Suedekum, J., & Woessner, N. (2017). German robots—The impact of industrial robots on workers. IAB-Discussion Paper No. 30.

Dehnbostel, P. (2018). Lern- und kompetenzförderliche arbeitsgestaltung in der digitalisierten arbeitswelt. *Arbeit, 27*(4), 269–294. https://doi.org/10.1515/arbeit-2018-0022

Deming, D. J. (2017). The growing importance of social skills in the labor market*. *The Quarterly Journal of Economics, 132*(4), 1593–1640. https://doi.org/10.1093/qje/qjx022

Dengler, K., & Matthes, B. (2018). The impacts of digital transformation on the labour market: Substitution potentials of occupations in Germany. *Technological Forecasting and Social Change, 137*, 304–316. https://doi.org/10.1016/j.techfore.2018.09.024

Dutta, S. (2010). What's your personal social media strategy? *Harvard Business Review, 88*(11):127–130, 151.

de Vries, G. J., Gentile, E., Miroudot, S., & Wacker, K. M. (2020). The rise of robots and the fall of routine jobs. *Labour Economics, 66*. https://doi.org/10.1016/j.labeco.2020.101885

Eggenberger, C., Rinawi, M., & Backes-Gellner, U. (2018). Occupational specificity: A new measurement based on training curricula and its effect on labor market outcomes. *Labour Economics, 51*, 97–107. https://doi.org/10.1016/j.labeco.2017.11.010

Eisele, S., & Schneider, M. R. (2020). What do unions do to work design? Computer use, union presence, and Tayloristic jobs in Britain. *Industrial Relations: A Journal of Economy and Society, 59*(4), 604–626. https://doi.org/10.1111/irel.12266

Eraut, M. (2004). Informal learning in the workplace. *Studies in Continuing Education, 26*(2), 247–273. https://doi.org/10.1080/158037042000225245

Feldmann, H. (2013). Technological unemployment in industrial countries. *Journal of Evolutionary Economics, 23*(5), 1099–1126. https://doi.org/10.1007/s00191-013-0308-6

Frey, C. B., & Osborne, M. A. (2017). The future of employment: How susceptible are jobs to computerisation? *Technological Forecasting and Social Change, 114*, 254–280. https://doi.org/10.1016/j.techfore.2016.08.019

Gardberg, M., Heyman, F., Norbäck, P. J., & Persson, L. (2020). Digitization-based automation and occupational dynamics. *Economics Letters, 189*. https://doi.org/10.1016/j.econlet.2020.109032

Geel, R., & Backes-Gellner, U. (2011). Occupational mobility within and between skill clusters: An empirical analysis based on the skill-weights approach. *Empirical Research in Vocational Education and Training, 3*(1), 21–38. https://doi.org/10.1007/BF03546496

Geel, R., Mure, J., & Backes-Gellner, U. (2011). Specificity of occupational training and occupational mobility: An empirical study based on Lazear's skill-weights approach. *Education Economics, 19*(5), 519–535. https://doi.org/10.1080/09645291003726483

Goos, M., Manning, A., & Salomons, A. (2009). Job polarization in Europe. *American Economic Review, 99*(2), 58–63. https://doi.org/10.1257/aer.99.2.58

Goos, M., Manning, A., & Salomons, A. (2014). Explaining job polarization: Routine-biased technological change and offshoring. *American Economic Review, 104*(8), 2509–2526. https://doi.org/10.1257/aer.104.8.2509

Haslberger, M. (2021). Routine-biased technological change does not always lead to polarisation: Evidence from 10 OECD countries, 1995–2013. *Research in Social Stratification and Mobility, 74*. https://doi.org/10.1016/j.rssm.2021.100623

Hellweg, T. (2021). Do employees with specific skill profiles receive more employer-funded training during technological change? Evidence from employer-employee data. In *19th ILERA World Congress*, Lund, Sweden.

Heß, P., Janssen, S., & Leber, U. (2019). *Digitalisierung und berufliche weiterbildung: Beschäftigte, deren tätigkeiten durch technologien ersetzbar sind, bilden sich seltener weiter*, Technical report, IAB-Kurzbericht.

Katz, L. F., & Murphy, K. M. (1992). Changes in relative wages, 1963–1987: Supply and demand factors. *The Quarterly Journal of Economics, 107*(1), 35–78. https://doi.org/10.2307/2118323

Lamo, A., Messina, J., & Wasmer, E. (2011). Are specific skills an obstacle to labor market adjustment? *Labour Economics, 18*(2), 240–256. https://doi.org/10.1016/j.labeco.2010.09.006

Lazear, E. P. (2009). Firm-specific human capital: A skill-weights approach. *Journal of Political Economy, 117*(5), 914–940. https://doi.org/10.1086/648671

Lindgren, H., & Schultze. (2004). Design principles for competence management systems: A synthesis of an action research study. *MIS Quarterly, 28*(3), 435–472. https://doi.org/10.2307/25148646

Lukowski, F., Baum, M., & Mohr, S. (2020). Technology, tasks and training—Evidence on the provision of employer-provided training in times of technological change in Germany. *Studies in Continuing Education, 48*, 1–22. https://doi.org/10.1080/0158037X.2020.1759525

Machin, S., & van Reenen, J. (1998). Technology and changes in skill structure: Evidence from seven OECD countries. *The Quarterly Journal of Economics, 113*(4), 1215–1244. https://doi.org/10.1162/003355398555883

Madia, S. A. (2011). Best practices for using social media as a recruitment strategy. *Strategic HR Review, 10*(6), 19–24. https://doi.org/10.1108/14754391111172788

Miranda, S., Orciuoli, F., Loia, V., & Sampson, D. (2017). An ontology-based model for competence management. *Data & Knowledge Engineering, 107*, 51–66. https://doi.org/10.1016/j.datak.2016.12.001

Mohr, S., Troltsch, K., & Gerhards, C. (2016). Job tasks and the participation of low-skilled employees in employer-provided continuing training in Germany. *Journal of Education and Work, 29*(5), 562–583. https://doi.org/10.1080/13639080.2015.1024640

Nedelkoska, L. (2013). Occupations at risk: Job tasks, job security, and wages. *Industrial and Corporate Change, 22*(6), 1587–1628. https://doi.org/10.1093/icc/dtt002

North, K., Reinhardt, K., & Sieber-Suter, B. (2018). Kompetenzmanagement in der Praxis: Mitarbeiterkompetenzen systematisch identifizieren, nutzen und entwickeln. Mit vielen Praxisbeispielen (3rd edn.) Wiesbaden: Springer Fachmedien Wiesbaden. https://doi.org/10.1007/978-3-658-16872-8

OECD. (2019). Education and training. In OECD (Ed.), *Measuring the digital transformation: A roadmap for the future* (pp. 172–173). OECD Publishing.

OECD. (2021). *Continuing education and training in Germany*. OECD Publishing.

Ore, O., & Sposato, M. (2021). Opportunities and risks of artificial intelligence in recruitment and selection. *International Journal of Organizational Analysis*. https://doi.org/10.1108/IJOA-07-2020-2291

Ouadahi, J. (2008). A qualitative analysis of factors associated with user acceptance and rejection of a new workplace information system in the public sector: A conceptual model. *Canadian Journal of Administrative Sciences/Revue Canadienne des Sciences de l'Administration, 25*(3), 201–213. https://doi.org/10.1002/cjas.65

Sebastian, R., & Biagi, F. (2018). The routine biased technical change hypothesis: A critical review. *Joint Research Centre (Seville site), European Commission*.

Seyda, S., Meinhard, D. B., & Placke, B. (2018). Weiterbildung 4.0 - digitalisierung als treiber und innovator betrieblicher weiterbildung. IW-Trends - Vierteljahresschrift zur empirischen Wirtschaftsforschung, *45*(1):107–124.

Spitz-Oener, A. (2006). Technical change, job tasks, and rising educational demands: Looking outside the wage structure. *Journal of Labor Economics, 24*(2), 235–270. https://doi.org/10.1086/499972

Venkatesh, V., Speier, C., & Morris, M. G. (2002). User acceptance enablers in individual decision making about technology: Toward an integrated model. *Decision Sciences, 33*(2), 297–316. https://doi.org/10.1111/j.1540-5915.2002.tb01646.x

Digital Twins in Flexible Online Work: Crowdworkers on German-Language Platforms

Paul Hemsen, Mareike Reimann, and Martin Schneider

Abstract External crowdworking (CW) is paid online work mediated by specialised crowdsourcing platforms. This chapter provides an introduction to various aspects of crowdworking with a focus on German-language platforms, based on the literature and our own results from the 'Digital Future' research programme. We define CW as an employment relationship and distinguish it from other forms of (non-)regular employment. Findings from a survey among crowdworkers show that crowdworkers are heterogeneous in terms of socio-demographic characteristics, and that the consequences of CW for health and work-life balance are ambivalent. Various platforms that broker complex tasks have developed a new type of rating system that commits workers to the platform. Based on crowdworkers' past performance record, they achieve a particular status level, such as 'five stars', which indicates a worker's reputation and determines the pay they can expect, as well as the tasks they can take on. Such rating-based compensation systems rely on a digital twin of each crowdworker that is stored by the platform. Today, such systems are platform-specific and proprietary, with a possible lock-in effect for employees. Public rating systems that cover multiple platforms are an alternative that would enable workers to transfer their reputation to other platforms. Overall, this chapter sheds light on an important but still under-researched form of flexible online work and illustrates that a novel form of the human digital twin is at the heart of platform management, with controversial implications for workers.

Keywords Crowdworkers · Temporary work · Rating systems

P. Hemsen · M. Schneider
Chair of Personnel Economics, University Paderborn, Warburger Str. 100, 33098 Paderborn, Germany
e-mail: paul.hemsen@uni-paderborn.de

M. Schneider
e-mail: martin.schneider@uni-paderborn.de

M. Reimann (✉)
Faculty of Sociology, Bielefeld University, Universitätsstraße 25, 33615 Bielefeld, Germany
e-mail: mareike.reimann@uni-bielefeld.de

© The Author(s), under exclusive license to Springer Nature Switzerland AG 2023
I. Gräßler et al. (eds.), *The Digital Twin of Humans*,
https://doi.org/10.1007/978-3-031-26104-6_12

1   Introduction

This chapter focuses on paid online work on public crowdworking platforms, which is called 'external crowdworking' (Giard et al., 2019). In this flexible form of work, various specialised online platforms, such as the microtask platform Amazon Mechanical Turk or the graphic design platform 99Designs, post online work tasks on behalf of their clients. Registered users—the crowdworkers—can decide to apply to perform any of these digital tasks and are paid after the satisfactory completion of a task by the intermediary CW platform.

CW has been discussed as a growing phenomenon of the worldwide digital economy (Boudreau et al., 2011; Horton & Chilton, 2010; Kässi & Lehdonvirta, 2018; Kittur et al., 2013; Fabo et al., 2017). In recent years, the opportunities and risks of CW have become an issue in a German context as well (Mrass et al., 2020; Pongratz & Bormann, 2017). Although CW seems to offer great possibilities for flexible work for diverse groups in the labour market (Reimann & Abendroth, 2023; Zyskowski et al., 2015), researchers, as well as union representatives, have also voiced concerns about labour protection (Barth & Fuß, 2021; Berg, 2016; De Stefano, 2016) and negative consequences such as health issues (Schlicher et al., 2021). Therefore, first, this chapter provides an introduction to external CW as a part of the rising gig economy (Kenney & Zysman, 2016). We review the relevant parts of the literature on CW and focus on the key findings of CW concerning German-language platforms. Using insights from a socio-economic perspective that interprets CW as a complex employment relationship, we compare CW with other forms of (non-)standard employment. Moreover, we present new empirical findings based on our own research. As part of the research programme 'Digital Future', which was run jointly by Bielefeld University and Paderborn University, interdisciplinary researchers conducted a German CW survey with 803 crowdworkers on four German-language CW platforms. The goal was to shed light on working conditions in digitalised work processes from the perspectives of computer scientists, economists, engineers, psychologists, and sociologists (for detailed information on the survey, see Giard et al. (2019, 2021)).

Next, we argue that a digital twin of each crowdworker is at the heart of a novel and sophisticated rating system used by multiple CW platforms. Once users have registered, platforms store socio-demographic information, along with the users' work histories. Crowdworkers also receive an overall rating (such as 'five stars'). The rating is typically based on performance evaluations by clients and on additional data stored by the platform (e.g., work history, processing time). Although the status hierarchies produced by these rating systems have been discussed before (Javadi Khasraghi & Aghaie, 2014; Goes et al., 2016; Goh et al., 2017), we add new insights to this literature. We show that among German-language CW platforms, these rating systems are crucial only for platforms that mediate complex tasks such as designing, testing, programming, or producing text. We also show how the digital twin of a crowdworker functions as the key element in these rating systems, helping platforms to match sophisticated tasks with qualified experts and to motivate and retain their crowdworkers. Hence, the digital twin determines the tasks that will be available to a crowdworker and the income they may be able to generate.

Overall, the chapter makes several contributions. We offer a socio-economic perspective on CW that considers employment relationships. For this purpose, we also report and discuss partially new empirical evidence on crowdworkers' characteristics, as well as possible opportunities and risks for work-life balance and health, with a particular focus on German-language platforms. Furthermore, we discuss the digital twins of workers in the context of external CW. This is an important issue. In a digitalised world of work, more information on employees will be assembled, and the resulting digital twins will influence employees' careers. CW is an environment in which this future development can be studied today.

In Sects. 2 to 4 of this chapter, we introduce external CW and distinguish it from other forms of flexible work, characterise CW as a new form of work, and report on empirical findings that indicate the strong heterogeneity of crowdworkers. Section 5 introduces the rating systems that platforms operate, examines how these systems are used by the platforms to allocate tasks and to provide incentives, and argues that there is a conflict of interest between platforms and workers concerning the public availability of the digital twins stored in the rating systems. Section 6 concludes by summarising our main points and projecting them onto future forms of flexible work.

2   External Crowdworking as Part of the Gig Economy

> Remember outsourcing? Sending jobs to India and China is so 2003. The new pool of cheap labor: everyday people using their spare cycles to create content, solve problems, even do corporate R & D. [...]. It's not outsourcing; it's crowdsourcing (Howe, 2006).

Jeff Howe introduced the term 'crowdsourcing' in 2006. It did not take long for academic and public discourse to focus on paid activities on crowdsourcing platforms and for the slightly modified term 'crowdworking' to gain prominence. CW has since become a growing phenomenon of the worldwide digital economy (Boudreau et al., 2011; Horton & Chilton, 2010; Kässi & Lehdonvirta, 2018; Kittur et al., 2013; Fabo et al., 2017; Taylor & Joshi, 2019), and this is also the case in a German context (Mrass et al., 2020; Pongratz & Bormann, 2017). In 2013, an estimated 48 million crowdworkers took on tasks brokered by CW platforms, and that figure was expected to rise to over 100 million crowdworkers in 2020, with an estimated gross service revenue of up to $25 billion (Kuek et al., 2015), making CW an important labour market (Schulte et al., 2020). In the European region, an estimated 9.2% of European and 6.9% of German workers are active crowdworkers or have at least practiced some kind of CW in the past (Serfling, 2019). Although there are no official figures on how many crowdworkers there really are, CW certainly provides interesting employment opportunities for people all over the world (Bracha & Burke, 2016). It attracts a wide range of people, from freelancers to employees in regular jobs, from people on parental leave to physically challenged people, and from students to retirees (Brabham, 2012).

Various terms are used to refer to work mediated by internet platforms, including 'crowdsourcing', 'gig economy', and 'platform economy'. Compared to these terms, CW differs in terms of the types of tasks offered, compensation, and contractual obligations (Schulte et al., 2020). First, CW platforms only involve digital tasks with digital outcomes (Schulte et al., 2020), which makes them globally accessible. The gig and platform economies also include locally restricted work, such as delivery, transportation, and craft services, which is mediated online but performed on site (Schulte et al., 2020). Digital tasks in CW vary from very simple and repetitive tasks that require only basic knowledge of how to use technical devices, such as tagging photos, answering surveys, or training artificial intelligence software, to sophisticated tasks such as writing text, producing graphic designs, or programming software (Durward et al., 2016). Generally, each platform deals with only a few related task types, which allows specialised CW platforms to connect specific clients with suitable experts. Second, CW is a subset of crowdsourcing that refers only to exchanges in which crowdworkers receive financial compensation for a contribution that is found to be satisfactory (Schulte et al., 2020).

Another important distinction is between the external and internal forms of CW. When CW is mentioned, it usually refers to what is known as external CW. External CW consists of task-based online work that is mediated through internet platforms. Internal CW, by contrast, refers to CW platforms created by a company with the intention of using the 'wisdom of the crowd' by using the participation of its own employees (plus external contributors in some cases) to solve mostly internal company challenges (Zuchowski et al., 2016; Abendroth et al., 2020). External CW is based on a triangular exchange involving the platform, clients, and crowdworkers (see also Sect. 4). The clients, who may be individuals, groups, or organisations, propose a digital task with a well-defined goal on an external CW platform (Estellés-Arolas & González-Ladrón-de Guevara, 2012). The platform displays these tasks online in the form of a call to a specified crowd, usually the platform's registered online users. The call includes descriptions of tasks and information about the benefits for each party involved. Like freelancers, crowdworkers undertake these tasks voluntarily, primarily on a task-by-task basis (Estellés-Arolas & González-Ladrón-de Guevara, 2012). They contribute their resources, such as time, money, effort, or expertise, and receive a range of benefits in return, such as intrinsic enjoyment of the activity and task-based payment (Estellés-Arolas & González-Ladrón-de Guevara, 2012). The use of such platforms is free for crowdworkers, but it is indirectly priced, as the platform retains a part of the task price paid by the client for providing a work environment, mediating between the crowdworker and client, and acting as a trustee.

3 Platforms in Germany: Expert Tasks Versus Microtasks

There is a surprisingly large number of CW platforms with German-language web interfaces and task offerings. Based on the findings by Hemsen (2021b), Table 1 lists 32 such platforms that fall into five commonly applied categories, and Fig. 1 compares the categories in detail, using characteristics that are often discussed in the literature.

The average German-language CW platform has about 100,000 (median) registered crowdworkers, with a strong variation depending on the task types that the platform offers. Among the 32 platforms, there are 8 design platforms (graphic design tasks), 6 market platforms (broader freelance tasks), 7 microtask platforms (simple and short tasks), 7 testing platforms (testing cases for software), and 4 text creation platforms (text writing tasks). On 24 of the 32 CW platforms, clients outsource tasks with a high complexity and a low granularity. In other words, on these platforms, crowdworkers generate holistic solutions for the platforms' clients. Therefore, it is not surprising that 22 of the 32 CW platforms conduct the qualification-based selection of crowdworkers and hold first-solution contests in which one or only a few crowdworkers work exclusively on a task to generate solutions. In this type of contest, the task price is paid for each acceptable solution. In contrast, in best-solution contests, crowdworkers work on a task simultaneously, and only the workers who create the best solutions are compensated. In addition, the context-specific selection of crowdworkers according to age, gender, or technical devices seems to be primarily relevant on test platforms. Surprisingly, given the large number of platforms for more complex tasks and the obvious need for experts with specific skills, only 8 of the platforms studied offer a rating-based compensation system (RBCS) in which pay and other intangible benefits are contingent on the crowdworker's rating or reputation, as certified by the platform. The logic of this rating system is examined more fully in Sect. 5 of this chapter. Overall, most German-language CW platforms require a high level of expertise and have created an appropriate working environment for this expertise. Microtask platforms are an exception.

Table 1 Overview of 32 German-language CW platforms

Platfrom type	German-language CW platforms
Designing platform	99designs; Brandsupply; Crowdsite; DesignCrowd; Designenlassen; jovoto; Logoarena; Talenthouse
Market platform	Bluepatent; Expertcloud; Fiverr; Freelancer; Twago; VoiceBunny
Microtask platform	Appjobber; Clickworker; Crowdguru; Gprofit; Streetspotr; Veuro; WorkGenius
Testing platform	Applause; Rapidusertest; test.io; testbee; Testbirds; Testemit.de; Uinspect; Testemit.de; Uinspect
Text creation platform	Content.de; Textbroker; Textmaster; Tripsbytips

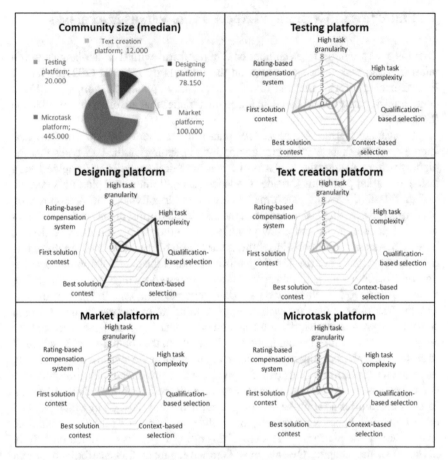

Fig. 1 Five platform types distinguished according to 9 platform characteristics, based on the 32 German-language CW platforms. *Note* Figure based on results from Hemsen (2021a). The highest value in the net diagram shows how many platforms belong to the platform type, and the thick black lines show how many of these platforms have a particular characteristic

Microtask platforms mainly offer tasks with a high granularity and low complexity. In other words, they are often repetitive, and it is difficult for crowdworkers to identify the underlying purpose of the task. Examples of these tasks are tagging photos or videos, answering surveys, maintaining product data, and performing simple research activities such as clarifying store hours. These tasks are usually taken on for the purpose of making money. However, given the simplicity of the tasks, they often offer low pay, and even with a 40-hour work week, crowdworkers on microtask platforms are unlikely to earn a four-figure monthly income (Giard et al., 2019). Since reputation systems such as RBCSs are rarely used on microtask platforms to improve working terms and conditions, including income, these remain unchanged regardless of the platform affiliation, performance, and behaviour of crowdworkers.

It is unclear why some CW platforms for demanding tasks have not implemented an RBCS, given the strong incentive and commitment effects (for a detailed discussion, see Sect. 5).

Based on what we know about German-language CW platforms, the widespread view, which considers crowdworkers to be merely digital day labourers for the simplest tasks, appears to be unfounded for most platforms, except perhaps for those that mediate microtasks. For German-speaking people, therefore, CW can be more than a type of productive pastime or a poor substitute for access to the 'regular' labour market. Instead, compared to regular employment, it is a labour market segment that is much more accessible to multiple groups of differently qualified people from virtually anywhere that only requires internet access.

4   The Crowdworkers

4.1   No Employment Contracts But Employment Relationships

CW can be understood as a new phenomenon of digital platform work, which in turn is part of the rising gig economy that became more prominent at the beginning of the 21st century (Kenney & Zysman, 2016). Nonetheless, it is comparable to other, more established forms of employment when it comes to important dimensions of the employment relationship. In this section, we will a) further explain why CW can be referred to as an employment relationship, and b) describe similarities to other forms of employment, namely regular employment, temporary work, (solo) self-employment, and fixed-term employment. Despite an increase in non-standard employment, the core of the German labour market is still characterised by regular employment relationships (Eichhorst & Tobsch, 2015). Thus, we will use regular employment, which refers to open-ended full-time employment contracts with a fixed employer, as a reference for comparisons, although CW is more similar to other forms of non-standard employment.

However, if employment relationships are understood to be more than a written employment contract, it becomes obvious that CW shares important features with other forms of employment, and systematisation can make it easier to understand the possible risks and benefits of CW. We consider the idea that employment relationships are more than just a legal employment contract and also more than the exchange of labour and money. Referring to the basic assumption of social exchange (Blau, 1964), they can be understood as multidimensional social exchange relationships (Brose et al., 2004; Coyle-Shapiro & Conway, 2004; Coyle-Shapiro & Kessler, 2000).

In regular employment and in most forms of non-standard employment, the partners involved in the exchange are an employee and an employer (Cappelli & Keller, 2013), and the exchange is embedded in an organisation that provides structural conditions and formal rules in a relatively stable context. Crowdworking is different. At its core is a relationship between the crowdworker and some client that is short-term and limited to the fulfilment of a very specific task, and thus the exchange is a clearly defined contribution for a (mostly) pre-determined payment. The crowdworker agrees to previously defined conditions for payment and task fulfilment by accepting the task. In focusing on single tasks, CW is very similar to (solo) self-employment; self-employed workers have exchange relationships with varying customers, although these exchanges can also be long-term, whereas CW tasks are typically short-term. The client mostly controls the definition of the content, time frame, and payment of a task, and finally decides to accept or not to accept the output of the crowd-worker. The crowdworker, in contrast, has limited control because they can only choose from tasks that are available on the platform. Communication with the client is mostly formalised and standardised, and direct interaction is often not possible.

However, CW differs from regular employment in another important way. It is a triangular relationship involving the crowdworker, the client, and the platform, which acts as an intermediary (Langley & Leyshon, 2017). Basically, the platforms mediate the demands of the client and the crowdworkers, who offer their time, competencies, and knowledge. The platform provides the environment, as well as the rules and conditions of the exchange, for both crowdworkers and clients. Crowdworkers accept the terms and conditions when they register on the platform (De Stefano, 2016). The platform's conditions stay the same for different tasks and different clients. Therefore, a series of exchanges is embedded into a relationship with the platform, which is more long-term. In its intermediary position, the platform is the most visible contact for crowdworkers, while the clients are often anonymous and invisible (De Stefano, 2016). Therefore, it can be assumed that the core of the exchange relationship is shifted from the employer-employee relationship to the crowdworker-platform relationship. Crowdworkers cannot negotiate contractual terms with the platform or with the clients, so the triangular relationship is defined by a power imbalance that puts the crowdworker at a disadvantage (Greef et al., 2020). Regarding this triangular relationship, CW is comparable to temporary work for which the hiring firm initiates and maintains a relationship between the lending firm and the employee.

It may be noted that crowdworkers might have some power because the plat-forms and the clients are reliant on the crowdworkers to work on the platforms, and thus, crowdworkers might advance their interests by threatening to leave the platforms. However, this is far from the daily reality of CW. Especially for crowd-workers on platforms for simple and repetitive tasks, i.e., microtasks, this possibility is excluded by the sheer mass of (globally) available potential crowdworkers with similar skills and knowledge. Crowdworkers on platforms for microtasks are there-fore easily replaceable (Howcroft & Bergvall-Kåreborn, 2019). However, it is true that platforms for more demanding tasks rely on a smaller number of qualified and willing crowdworkers than platforms for microtasks (Schulten & Schaefer, 2015; Boons et al., 2015). A valuable tool for retaining qualified crowdworkers can be rating-based compensation systems (see Sect. 5). Again, however, crowdworkers are not in a position to negotiate. There are still two options: Take it or leave it.

It is useful to take a closer look at how the content of the employment relation-ship in CW differs from, or is sometimes similar to, the content of this relationship in other forms of employment. We will focus on some important dimensions of this relationship, namely job security, earnings, social security contributions, and flexibility.

Job security: In CW, similar to (solo) self-employment, there is no job security, as there is no employment contract that defines the duration of the employment relationship. The duration of the exchange is limited to the fulfilment of a specified task, with no promise of further work opportunities. In comparison, job security is highest in regular employment because of the permanent, open-ended contract and corresponding legal protection against dismissal. The duration is also determined in fixed-term employment, as there is a fixed ending point to the exchange that is set in advance, although in Germany fixed-term contracts are often transferred to permanent contracts, at least in specific industries or occupations (Eichhorst & Tobsch, 2015). Moreover, fixed-term contracts involve a longer time frame compared to the short-term nature of CW. In temporary work, the contract with the hiring firm is limited and insecure, but temporary workers often have permanent contracts with the lending firms and thus a comparably stable employment situation.

Earnings Ideal-typical regular employment provides a living wage. This is also true for temporary work or fixed-term employment, although earnings are on average lower compared to regular employment (Giesecke & Verwiebe, 2009). Crowdworkers' earnings are highly task-dependent. However, studies have shown that possible as well as average earnings differ significantly between different platforms and specific task categories, and they depend on the necessary qualifications, experiences, or competencies, and on whether crowdworkers search for available tasks throughout the day on multiple platforms (Lehdonvirta, 2018). Therefore, earnings are highly dependent on individual engagement and also on the availability of (well-paid) tasks (Wood et al., 2019). Crowdworkers often have small hourly and unstable earnings, although there is a high variability in these earnings (Berg, 2016). As there is a low threshold for accessing CW, there is strong competition, and crowdworkers can be replaced easily compared to employees, limiting the amount of payment for tasks. As a result, higher earnings are rare (Hornuf & Vrankar, 2022), and CW is often used for additional income (Berg, 2016). In this regard, CW is again comparable to (solo) self-employment, where workers are dependent on demand. Some self-employed workers earn very little income, but others can compensate for this insecurity by generating very high incomes for their very specific qualifications or competencies (Hamilton, 2000).

Social security contributions: The social security system in Germany is based on a high share of regular employment, where employees are fully integrated into all social security systems through contributions by employers and employees (pension, health, unemployment). In temporary work, fixed-term employment, and part-time work, employees are also fully integrated into all social security systems, although their contributions are lower due to comparably lower income. Self-employed workers are fully responsible for taxes and social security contributions. They can pay into the systems on a voluntary basis, use private provisions, or not pay at all. In this case, CW is again comparable to self-employment, as crowdworkers are also fully responsible for paying social security contributions.

Flexibility: An important aspect that is discussed in the context of CW is flexibility, both from the perspective of the client and from the perspective of the crowdworker (Abendroth et al., 2020). In regular employment, both employer and employee flexibility are low due to strong employment security and the employment contract. Although this is a benefit in terms of job security, predictability, and planning capacity for both the employer and the employee, it restricts short-term reactions to changes in demand for the employer and causes inflexibility in the adaption to individual preferences and changes in the private life of the employee. In contrast, flexibility is high in CW on both the crowdworker side and the client side. Clients can quickly react to (short-term) demands for specific knowledge or competencies by outsourcing tasks to CW platforms and assessing a heterogeneous pool of crowdworkers (Leimeister et al., 2016). Crowdworkers are highly flexible because they can easily change platforms and tasks and because they can schedule their work around other responsibilities such as childcare, other employment, or individual preferences (Reimann & Abendroth, 2023; Warren, 2021). Additionally, working hours are not fixed, so CW makes it possible to earn more if this is necessary or preferred, and to change the amount of effort put into CW very flexibly, even on a day-to-day basis. However, this flexibility is highly dependent on the available tasks and changes in the demand for specific knowledge or competencies, and therefore again CW is very similar to (solo) self-employment and also to temporary work.

In summary, the platform-based mediation of tasks is a core element of CW. Even though crowdworkers do not have a written employment contract, they still enter employment relationships with characteristics that are comparable to the characteristics of other forms of employment. CW is especially similar to (solo) self-employment when it comes to task-based work assignment, the dependence on demand, uncertainties in earnings, the long-term perspective of work, social security contributions, and also the advantage of flexibility. CW is also similar to temporary work, especially in terms of the triangular relationship between crowdworkers, platforms, and clients.

4.2 The Diverse Crowd: Demographics, Health, and Work-Life Balance

Crowdworkers are as diverse as the platforms that they use. Research on CW in Germany suggests that overall crowdworkers are rather young, male, and well-educated individuals, but it has become increasingly obvious that CW is used by heterogeneous workers with various motivations, which we will discuss in this section.

There have been many studies on the (socio-demographic) characteristics of crowdworkers in Germany (e.g., Leimeister et al. (2016); Pesole et al. (2018); Serfling (2019); and our own interdisciplinary German CW survey (Giard et al., 2019, 2021)). Before going into detail on specific aspects, it should be noted that existing studies are only partly comparable because the study designs differ markedly. As there is no obligatory registration of crowdworkers in Germany, it is not possible to draw representative samples of the crowdworker population. As a result, researchers have used various strategies to collect data about CW; they mostly use self-information online questionnaires, but they also use telephone interviews or algorithm-based data collection on crowdworker behaviour. Consequently, there are similarities between studies and results, but also differences due to the survey methods used, the included platforms, and even the underlying definition of CW (Abendroth et al., 2020; Giard et al., 2019).

Existing surveys performed in Germany, including our own, are somewhat consistent in terms of the mean age of crowdworkers: Crowdworkers are on average about 36 years old (mean values ranging between 36.8 years old in Giard et al. (2021) and 35.6 years old in Leimeister et al. (2016)). However, the average age varies between platform types: On the testing platforms, crowdworkers are on average 32.8 years old; on microtask platforms, they are 36.6 years old; and on marketplace platforms, they are 43.8 years old. Overall, crowdworkers seem to be younger than the average working population in Germany (mean age of 44.1 years old in 2019; BiB, 2019). There are more male than female crowdworkers across all studies (Berg, 2016; Leimeister et al., 2016; Serfling, 2019), with a ratio of roughly 60/40. However, analyses based on our interdisciplinary German CW survey show that the gender distribution differs significantly between platform types: There is a higher share of males on testing (69%) and mobile CW platforms (68%), but there is a higher share of female crowdworkers on marketplace platforms (61%) (Giard et al., 2021). On average, crowdworkers are highly educated, with most of them having 'Abitur' (the highest level of general school education, which is necessary for university entrance) or a tertiary degree (Giard et al., 2021; Leimeister et al., 2016). The motives for performing CW are also diverse. They can be intrinsic motives such as self-fulfilment, fun, content-related interest in the tasks, or the advancement of individual competencies and experiences, but they can also be extrinsic motives such as earning money and acquiring new customers as a self-employed worker (Al-Ani & Stumpp, 2015; Feldmann et al., 2018).

In addition to studying the characteristics and motivations of crowdworkers, scholars have increasingly investigated the consequences of CW for different aspects of life, such as work-life balance and health.

Work-life balance

The flexibility involved in CW makes it possible for individuals to facilitate work-life balance. In research on the work-life interface, flexible working is referred to as a job resource that allows for the more individual scheduling of work obligations and thus facilitates the integration of work and a person's private life (Hill et al., 2010; Schieman & Glavin, 2008). In line with this, work-family border theory (Clark, 2000) and boundary management theory (Kossek, 2016) specify that having control over one's schedule allows for flexible adaptation and the coordination of the timing of work demands with private obligations.

CW can be accessible for almost everyone because it offers low barriers for flexible working. It involves a high potential flexibility (Reimann & Abendroth, 2023): Crowdworkers have easy access to job tasks via platforms, and they can perform the job tasks at the location of their choice because tasks can be done completely digitally using internet-ready devices such as computers and smartphones (Berg, 2016). For example, CW can be conducted at home but also at a café or while commuting to another job and thus has a high degree of flexibility in terms of location. CW can be carried out on any day of the week and at any time of day (De Stefano, 2016). Work schedules and the lengths of individual tasks can vary depending on the crowdworker's selection of work tasks (Pesole et al., 2018). Consequently, CW is characterised by a high degree of flexibility in time as well: The timing of the beginning and end of work, the scheduling of breaks and days off, and the distribution of work over the day or week can easily be adjusted so that they do not conflict with obligations from other jobs or from private life. Finally, CW is characterised by a high degree of task autonomy, because crowdworkers control which tasks they choose from the available tasks. They may decide to do more or less complex or time-consuming tasks, tasks that seem more interesting than others, tasks that are more appropriate for their individual skills and knowledge, or tasks that have a better anticipated cost-benefit ratio than others (Howcroft & Bergvall-Kåreborn, 2019).

However, permanent switching between roles requires complex individual self-management abilities and may also result in an increased blurring of boundaries (Kossek et al., 2006). Thus, despite the high potential flexibility of CW, the actual daily life of a crowdworker may look somewhat different (Lehdonvirta, 2018), and exposure to employment insecurity and market pressures might counteract flexibility (De Stefano, 2016). Formal autonomy may come at the price of long, unsocial, and irregular working hours (Berg et al., 2018), and the need to constantly check for available tasks if a decent income needs to be earned (Wood et al., 2019); this may have negative effects on work-life integration.

Our own research based on the interdisciplinary German CW survey shows that flex-time and flex-place in CW are not as widespread among crowdworkers as this type of digital labour would seem to suggest. However, if crowdworkers do work flexibly in time and if they have high task autonomy, they are indeed able to benefit from flexible working hours, as they experience fewer work-life conflicts. This does not apply to working flexibly in place. In contrast, if crowdworkers are not able to choose their tasks autonomously or if they are restricted in terms of when they perform their tasks, CW increases the likelihood that work strain will seep into non-work life (Reimann & Abendroth, 2023).

Health

Although CW has been discussed as an opportunity for workers with health issues (Zyskowski et al., 2015), little is known about the possible health consequences of CW. A long tradition of research on non-standard work arrangements, however, has already shown their possibly negative impact on an individual's perceived stress, mental health, musculoskeletal problems, and other physical health problems (e.g., Quinlan et al. (2001); Sverke et al. (2002)). As CW can be compared to those forms of employment (see Sect. 4.1), its health risks may be similar. Research on technology-enabled work indicates possible mental and physical health risks as well. Examples of these risks are isolation and a lack of support when working from home (Collins et al., 2016; Cooper & Kurland, 2002; Tavares, 2017), as well as technology-driven work intensification (Meyer et al., 2019). As CW is completely internet-based and carried out on computers or smartphones, the risks of digitalised work and working from home are relevant for CW as well. Moreover, irregular and unsocial working hours, which seem to be very common in CW (Berg et al., 2018), are associated with impaired health as well (Costa et al., 2006).

In research based on the interdisciplinary German CW survey, we analysed whether participation in CW is linked to increased somatic symptoms compared to regularly employed personnel. We found that crowdworkers show significantly increased somatic symptoms compared to a German norm sample. The higher symptoms are stable across different kinds of tasks and platforms, gender, and age groups, and they are statistically related to the extent of participation in CW. Specifically, we found that the total work hours per week were not associated with an increase in somatic symptoms, but we did find associations with strain-based work-family conflict and with earning money being the primary motivation to participate in CW (Schlicher et al., 2021).

5 Rating Systems on Platforms

5.1 An Instrument to Allocate Tasks, as Well as to Attract, Motivate, and Retain Crowdworkers

Rating systems are already common in online shopping, on social media platforms, and on a variety of other internet platforms; they are used, for example, to rate products or content from social media profiles (Jøsang et al., 2007). Not surprisingly, CW platforms have also developed rating systems to quantify crowdworkers' overall performance (Gandini, 2019). Figure 2 provides an example of an RBCS used by the German text creation platform Textbroker. In this example, crowdworkers receive one to five stars. Importantly, individual compensation and the attractiveness of the available tasks depend on this rating. The number of stars in an RBCS is typically influenced by the crowdworker's performance record in terms of the length of membership, number of tasks completed, and quality of task fulfilment (Hemsen, 2021b).

Rating systems are platform-specific and their details vary, for example, in terms of the importance of monetary incentives and the granularity of the ratings. Although the previous literature has not systematically explained how the more sophisticated rating systems of Textbroker and similar platforms work (more on this below), some evidence is available that suggests that the rating system is an important element in

Fig. 2 Example of an RBCS for a German-language CW platform for text creation. *Note* A star rating is the core element of the RBCS of the German text creation platform Textbroker (screenshot: 02.09.2021). In order to achieve a higher rating on a text creation platform, it is usually important that the produced texts are error-free in terms of grammar, spelling, punctuation, and expressions. The platform also checks whether or not the client's specifications have been met. Theoretically, every written text can improve the rating, but the rating is updated only over longer intervals, and the platforms also do not usually justify the amount of pay for each rating level

each platform's business model. Basically, the rating system is supposed to attract and commit crowdworkers to the platform by providing incentives. It is also meant to help platforms to match tasks with crowdworkers, thus increasing the quality of the completed tasks and client satisfaction.

In line with this view, rating systems have been shown to exert a positive influence on the performance and participation of crowdworkers on a platform (Schörpf et al., 2017). Even for rating systems without monetary incentives, users with higher rating levels will perform better, allegedly based on the displayed reputation (Peer et al., 2014; Basili & Rossi, 2020; Goes et al., 2016). The reputation also encourages crowdworkers to adapt their behaviour to the requirements of the platform and its clients (Riedl & Seidel, 2018). In addition, positive effects on crowdworker participation are achieved through virtual reward systems with gamified elements, such as ratings, that amplify intrinsic crowdworker motivation (Feng et al., 2018; Goh et al., 2017). Similarly, direct performance evaluations by clients and peers, even during an ongoing work process, also affect performance (Jian et al., 2019).

The literature does not discuss one important distinction. While some rating systems are based on reputation and fame alone, some are tied to considerable monetary incentives. We call these more sophisticated examples 'rating-based compensation systems' (RBCSs) (Hemsen, 2021a, b). RBCSs differ from simple 'status hierarchies' in which recognition evolves spontaneously within a community based on visible contributions and positive feedback (Goes et al., 2016). Instead, platforms carefully craft the ratings in RBCSs. Performance is measured with multiple criteria, including the number of tasks solved and their quality according to the subjective evaluations of the platform provider and the clients. Furthermore, the crowdworker category, which may be the number of stars or some achieved title, has important implications because the RBCSs combine material and immaterial rewards. More stars, badges, or titles imply more recognition or a higher reputation, as individual ratings are visible to peers and clients (Auriol & Renault, 2001; Goes et al., 2016). A higher rating level also leads to material rewards (Auriol & Renault, 2001). Crowdworkers with higher ratings typically receive higher pay rates per task, bonuses, and privileged access to more lucrative and interesting tasks. As Fig. 2 shows, at Textbroker, five-star workers receive considerably higher pay than four-star workers.

As a result, RBCSs are sophisticated instruments that simulate an incentive hierarchy or so-called 'internal labour market' within a traditional organisation, but without the contractual obligations of an employer-employee relationship (Hemsen, 2021a, b). Moving to a higher rating is comparable to a promotion (Auriol & Renault, 2001). The hierarchy of status levels describes a predefined career path for registered crowdworkers on the platform. In order to be promoted, they need to invest specifically by being active on a particular platform and by receiving favourable feedback from the platform and its clients. Hence, the effects of this system are also quite similar to those of internal labour markets (Hemsen, 2021a, b). This deters workers who are not interested in a more long-term engagement, and attracts workers who are. These latter crowdworkers become to some extent bound to the platform because leaving it causes a loss of reputation and the associated rewards. This is because the RBCS is platform-specific. On comparable platforms, workers would have to start at the bottom of the incentive hierarchy. This effect restricts crowdworkers' flexibility and mobility more the longer they have been registered and the higher they are in the incentive hierarchy. Hence, the RBCS not only commits crowdworkers to the platform but also stimulates the accumulation of platform-specific expertise, because crowdworkers often work on similar tasks over time and receive continuous performance feedback.

Empirical work using the interdisciplinary German CW survey provides some evidence on the commitment effects that the RBCSs exert. Platforms with an RBCS have significantly more committed crowdworkers who work more hours per week than crowdworkers on platforms with non-reputational fixed task prices (Hemsen, 2021a). In addition, both emotional and rational economically driven commitment to the platform are found to increase significantly with each rating level. To some extent, the number of hours worked per week also increases for higher ratings, provided that sufficiently high incentives are offered.

Although RBCSs are applicable to many different types of platforms, they are not that widespread in the German-language CW market. Among the 32 CW platforms surveyed (see Fig. 1 above), only 8 have implemented an RBCS—and these platforms broker more complex tasks. Platforms for complex tasks rely on the commitment and motivation of their expert crowdworkers to keep the business running, and this can be achieved by the RBCS. In contrast, microtask CW platforms can rely on the existing crowdworkers who are willing to take on microtasks from time to time, and each crowdworker can be replaced by others given the low level of expertise involved. Therefore, we interpret the RBCS as a sophisticated solution to the problem of retaining and incentivising expert workers on highly specialised platforms.

5.2 A Crowdworker's Record in the Rating System Is a Digital Twin

RBCSs or other forms of monetary or non-monetary reputation systems on CW platforms create a digital twin for each crowdworker. In this context, the digital twin consists of all the information collected by the platform on a specific worker, including, for example, their age, sex, time of registration with the platform, qualifications, and skills, the number and level of tasks they have taken on, and the clients' quality ratings. The digital twin forms the basis for assigning the crowdworker to a level in the RBCS and hence for matching workers and tasks. Thus, it is crucial to the business models of many CW platforms. Even on microtask platforms, the digital twin is important. For example, a client on a microtask platform may want to conduct a survey on a particular topic and may only be interested in the opinion of a specific target group (e.g., women who are more than 40 years old), but also needs reliable participants within a short period of time. The digital twins of the crowdworkers make it possible to selectively offer the survey task to the specific target group and to those who have reliably fulfilled similar tasks within a preset time frame in the past. Another example is that a client on a designing platform might need crowdworkers with verifiable experience in designing labels for a specific product group, such as soft drinks, who also speak Italian. In this case, the digital twins are also used to select those crowdworkers who are most likely to fulfil the task.

The digital twins within the RBCS are important because without them platforms would have little information about their diverse crowdworkers; platforms would lack information about a crowdworker's actual performance level, motives, and personal background (Boons et al., 2015; Schulten & Schaefer, 2015). This lack of information is a problem, especially for CW platforms that rely on qualified experts who they match with clients who demand their expertise (Schulten & Schaefer, 2015). To attract, motivate, and retain such experts, platforms must be aware of the economic and social needs of their crowdworkers. This requires information about their crowdworkers. Crowdworkers, in turn, benefit from such systems by being able to satisfy their social need for recognition, status, or reputation, while platforms with an RBCS are also able to offer more desirable compensation that is more in line with the required skill level, as well as opportunities for incremental improvement in performance and behaviour.

One crucial effect of RBCSs is the commitment effect that they have on workers. As part of the 'Digital Future' research programme, Schneider and Hemsen (2021) demonstrated that crowdworkers developed different types and different degrees of commitment to a text creation platform, depending on their specific personal circumstances. Multiple distinct groups of crowdworkers were found to exhibit similar emotionally or rationally based commitments to the platform, and commitment was reflected in group-specific patterns of participation and the intention to stay. For example, the most important group of crowdworkers on the platform studied are strongly motivated by additional income and not interested in simply passing the time, and they are rationally committed to the platform. This group consists mostly of self-employed persons and persons who report CW as their main occupation. These findings support the commitment effect of RBCS—and they imply that platforms could make their incentive system even more attractive by tailoring their rewards specifically to these groups, thus benefitting all parties involved. The findings were based on survey data, which the platform could also request from workers to complement the digital twins. Hence, our findings illustrate more generally how the digital twins stored in rating systems can be used by platforms to learn about their diverse crowdworkers.

However, rating systems and the digital twin are platform-specific and proprietary. Crowdworkers are neither able nor allowed to transfer their reputation and status to other platforms or companies. As a result, the digital twin may be locked into the particular CW platform that created it, as there is no standardised way to merge information from different platforms. This has drawbacks for all parties involved. For crowdworkers, it leads to so-called vendor lock-in: No information about the qualifications or reputation of the crowdworker is shared between different platforms (Hemsen et al., 2020). Therefore, crowdworkers who leave a platform because they want to invest more time in a more lucrative or interesting platform simultaneously lose their reputation on the previous platform (Hemsen et al., 2020). Similarly, crowdworkers who work for multiple CW platforms, which is not uncommon, may be undervalued compared to their counterparts who work for only one platform. The consequence of having undervalued crowdworkers is that their skills may be underused. Of course, platforms might benefit in the short run from qualified crowdworkers who are locked in, but in the long term, the underutilisation of the skills of undervalued crowdworkers can negatively impact the quality of the solutions offered to clients (Hemsen et al., 2020).

Conceptual work by researchers from the 'Digital Future' research programme suggests that an appropriate solution consists of the platform-independent management and storage of crowdworker information (Hemsen et al., 2020). Such platform-independent management and storage systems can mitigate the effects of vendor lock-in, as crowdworkers can freely transfer and share their information and thus, their digital twin. By making crowdworker information available to all CW platforms to which a crowdworker has been granted access, platforms can reduce the cost of information collection, which is likely to improve the fit between crowdworkers and tasks, and thus client satisfaction. Whether platforms are willing to standardise and share information collected from crowdworkers, whether legislators may need to enforce this by law, and what a system for managing and storing crowdworker data should look like are still unclear and call for future research.

6 Conclusions

Although CW seems to involve highly flexible short-lived gigs at first glance, it has the potential to give rise to more long-term relationships. As an employment form, it shares important features with a number of other employment forms, namely regular employment, temporary work, and self-employment, but it is still is a unique and novel form of work arrangement. The incomes reported by German crowdworkers vary considerably, so CW should not as a whole be considered an exploitative form of day labour. Though CW tends to be compatible with a good work-life balance, some of its potential health effects are problematic. The rating system, which creates digital twins of crowdworkers, is a central element in CW (and, by implication, in other forms of flexible work). Crowdworkers are extremely diverse in terms of their ages, personal situations, and motives. Therefore, a platform can learn about its workers' expertise and commit workers to the platform by using sophisticated rating systems, which are based on creating crowdworker digital twins. Crowdworkers in turn rely on their reputation according to their digital twin to access interesting and lucrative work tasks. Today, rating systems are platform-specific and proprietary. There is already some discussion on public rating systems that cover various platforms. This discussion should continue and potentially include other forms of work, because employer-operated rating systems will become more comprehensive and have more of an influence on workers' careers.

References

Abendroth, A. K., Reimann, M., Diewald, M., & Lükemann, L. (2020). Arbeiten in der crowd: Perspektiven der theorie relationaler ungleichheiten in arbeitsorganisationen. *Industrielle Beziehungen Zeitschrift für Arbeit, Organisation und Management, 27*(2–2020), 160–178. https://doi.org/10.3224/indbez.v27i2.04

Al-Ani, A., & Stumpp, S. (2015). *Motivationen und Durchsetzung von Interessen auf kommerziellen Plattformen.* Ergebnisse einer Umfrage unter Kreativ- und IT-Crowdworkern: HIIG Discussion Paper Series. https://doi.org/10.2139/ssrn.2699065

Auriol, E., & Renault, R. (2001). Incentive hierarchies. Annales d'Économie et de Statistique (63/64), 261. https://doi.org/10.2307/20076305

Barth, V., & Fuß, R. (2021). Crowdwork und die aktivitäten der ig metall. *Zeitschrift für Arbeitswissenschaft, 75*(2), 182–186. https://doi.org/10.1007/s41449-021-00246-x

Basili, M., & Rossi, M. A. (2020). Platform-mediated reputation systems in the sharing economy and incentives to provide service quality: The case of ridesharing services. *Electronic Commerce Research and Applications, 39*. https://doi.org/10.1016/j.elerap.2019.100835

Berg, J. (2016). Income security in the on-demand economy: Findings and policy lessons from a survey of crowdworkers. *Comparative Labor Law and Policy Journal, 37*(3), 543–576.

Berg, J., Furrer, M., Harmon, E., Rani, U., & Six, S. M. (2018). *Digital labour platforms and the future of work: Towards decent work in the online world.* Geneva: International Labour Office.

BiB. (2019). Alterung und Arbeitsmarkt, 26.09.2019. Available at: https://www.bib.bund.de/DE/Presse/Konferenzen/2019-09-26-Alterung-und-Arbeitsmarkt.html

Blau, P. M. (1964). *Exchange and power in social life.* New York, NY: Wiley.

Boons, M., Stam, D., & Barkema, H. G. (2015). Feelings of pride and respect as drivers of ongoing member activity on crowdsourcing platforms. *Journal of Management Studies, 52*(6), 717–741. https://doi.org/10.1111/joms.12140

Boudreau, K. J., Lacetera, N., & Lakhani, K. R. (2011). Incentives and problem uncertainty in innovation contests: An empirical analysis. *Management Science, 57*(5), 843–863. https://doi.org/10.1287/mnsc.1110.1322

Brabham, D. C. (2012). The myth of amateur crowds. *Information, Communication & Society, 15*(3), 394–410. https://doi.org/10.1080/1369118X.2011.641991

Bracha, A., & Burke. M. A. (2016). Who counts as employed? Informal work, employment status, and labor market slack. *Federal Reserve Bank of Boston Working Papers, 16-29,* 1–42.

Brose HG, Diewald, M., & Goedicke, A. (2004). Arbeiten und haushalten. Wechselwirkungen zwischen betrieblichen beschäftigungspolitiken und privater lebensführung. Beschäftigungsstabilität im Wandel? (pp. 287–310).

Cappelli, P., & Keller, J. (2013). Classifying work in the new economy. *Academy of Management Review, 38*(4), 575–596. https://doi.org/10.5465/amr.2011.0302

Clark, S. C. (2000). Work/family border theory: A new theory of work/family balance. *Human Relations, 53*(6), 747–770. https://doi.org/10.1177/0018726700536001

Collins, A. M., Hislop, D., & Cartwright, S. (2016). Social support in the workplace between teleworkers, office-based colleagues and supervisors. *New Technology, Work and Employment, 31*(2), 161–175. https://doi.org/10.1111/ntwe.12065

Cooper, C. D., & Kurland, N. B. (2002). Telecommuting, professional isolation, and employee development in public and private organizations. *Journal of Organizational Behavior, 23*(4), 511–532. https://doi.org/10.1002/job.145

Costa, G., Sartori, S., & Akerstedt, T. (2006). Influence of flexibility and variability of working hours on health and well-being. *Chronobiology International, 23*(6), 1125–1137. https://doi.org/10.1080/07420520601087491

Coyle-Shapiro, J., & Kessler, I. (2000). Consequences of the psychological contract for the employment relationship: A large scale survey. *Journal of Management Studies, 37*(7), 904–930. https://doi.org/10.1111/1467-6486.00210

Coyle-Shapiro, J. A. M., & Conway, N. (2004). The employment relationship through the lens of social exchange. In J. A. M. Coyle-Shapiro, L. M. Shore, S. M. Taylor, & L. Tetrick (Eds.), *The employment relationship: Examining psychological and context perspectives* (pp. 5–28). Oxford, U.K.: Oxford University Press.

De Stefano, V. (2016). The rise of the "just-in time workforce: On demand work, crowdwork, and labor protection in the "gig economy. *Comparative Labor Law and Policy Journal, 37*(3), 471–504.

Durward, D., Blohm, I., & Leimeister, J. M. (2016). Crowd work. *Business & Information Systems Engineering, 58*(4), 281–286. https://doi.org/10.1007/s12599-016-0438-0

Eichhorst, W., & Tobsch, V. (2015). Not so standard anymore? Employment duality in Germany. *Journal for Labour Market Research, 48*(2), 81–95. https://doi.org/10.1007/s12651-015-0176-7

Estellés-Arolas, E., & González-Ladrón-de Guevara, F. (2012). Towards an integrated crowdsourcing definition. *Journal of Information Science, 38*(2), 189–200. https://doi.org/10.1177/0165551512437638

Fabo, B., Beblavý, M., Kilhoffer, Z., & Lenaerts, K. (2017). *An overview of European platforms: Scope and business models.* Luxembourg: Publications Office of the European Union.

Feldmann, C., Hemsen, P., & Giard, N. (2018). Crowdworking: Einflüsse der arbeitsbedingungen auf die motivation der crowd worker.

Feng, Y., Ye, H. J., Yu, Y., Yang, C., & Cui, T. (2018). Gamification artifacts and crowdsourcing participation: Examining the mediating role of intrinsic motivations. *Computers in Human Behavior, 81*, 124–136. https://doi.org/10.1016/j.chb.2017.12.018

Gandini, A. (2019). Labour process theory and the GIG economy. *Human Relations, 72*(6), 1039–1056. https://doi.org/10.1177/0018726718790002

Giard, N., Hemsen, P., Hesse, M., Löken, N., Nouri, Z., Reddehase, J., Reimann, M., Schlicher, K., & Schulte, J. (2019). Interdisziplinäre befragung von crowdworkern. Technical Report.

Giard, N., Brunsmeier, S., Hemsen, P., Hesse, M., Löken, N., Nouri, Z., Reimann, M., Schlicher, K., Schneider, M., & Schulte, J. (2021). Erkenntnisse zur arbeitsrealität deutscher crowdworker.

Giesecke, J., & Verwiebe, R. (2009). The changing wage distribution in Germany between 1985 and 2006. *Schmollers Jahrbuch, 129*(2), 191–201. https://doi.org/10.3790/schm.129.2.191

Goes, P. B., Guo, C., & Lin, M. (2016). Do incentive hierarchies induce user effort? Evidence from an online knowledge exchange. *Information Systems Research, 27*(3), 497–516. https://doi.org/10.1287/isre.2016.0635

Goh, D. H. L., Pe-Than, E. P. P., & Lee, C. S. (2017). Perceptions of virtual reward systems in crowdsourcing games. *Computers in Human Behavior, 70*, 365–374. https://doi.org/10.1016/j.chb.2017.01.006

Greef, S., Schroeder, W., & Sperling, H. J. (2020). Plattformökonomie und arbeitsbeziehungen: Digitalisierung zwischen imaginierter zukunft und empirischer gegenwart. *Industrielle Beziehungen Zeitschrift für Arbeit, Organisation und Management, 27*(2–2020), 205–226. https://doi.org/10.3224/indbez.v27i2.06

Hamilton, B. H. (2000). Does entrepreneurship pay? An empirical analysis of the returns to self-employment. *Journal of Political Economy, 108*(3), 604–631. https://doi.org/10.1086/262131

Hemsen, P. (2021a). How do rating-based compensation systems on crowdworking platforms work? *Providing long-term and reputational compensation for expert crowdworkers.*

Hemsen, P. (2021b). Rating-based compensation systems as a commitment tool on crowdworking platforms: An empirical analysis of four platforms.

Hemsen, P., Hesse, M., Löken, N., & Nouri, Z. (2020), Platform-independent reputation and qualification system for crowdwork. In *2nd Crowdworking Symposium.*

Hill, E. J., Erickson, J. J., Holmes, E. K., & Ferris, M. (2010). Workplace flexibility, work hours, and work-life conflict: Finding an extra day or two. *Journal of Family Psychology, 24*(3), 349–358. https://doi.org/10.1037/a0019282

Hornuf, L., & Vrankar, D. (2022). Hourly wages in crowdworking: A meta-analysis. *SSRN Electronic Journal.* https://doi.org/10.2139/ssrn.4020691

Horton, J. J., & Chilton, L. B. (2010). The labor economics of paid crowdsourcing. In D. C. Parkes, C. Dellarocas & M. Tennenholtz (Eds.), *Proceedings of the 11th ACM Conference on Electronic Commerce* (pp. 209–218). New York, NY, USA: ACM Press. https://doi.org/10.1145/1807342.1807376

Howcroft, D., & Bergvall-Kåreborn, B. (2019). A typology of crowdwork platforms. *Work, Employment and Society, 33*(1), 21–38. https://doi.org/10.1177/0950017018760136

Howe, J. (2006). The rise of crowdsourcing. *Wired Magazine, 14*(6), 1–5.

Javadi Khasraghi, H., & Aghaie, A. (2014). Crowdsourcing contests: Understanding the effect of competitors' participation history on their performance. *Behaviour & Information Technology, 33*(12), 1383–1395. https://doi.org/10.1080/0144929X.2014.883551

Jian, L., Yang, S., Ba, S., Lu, L., & Jiang, L. C. (2019). Managing the crowds: The effect of prize guarantees and in-process feedback on participation in crowdsourcing contests. *MIS Quarterly, 43*(1), 97–112. https://doi.org/10.25300/MISQ/2019/13649

Jøsang, A., Ismail, R., & Boyd, C. (2007). A survey of trust and reputation systems for online service provision. *Decision Support Systems, 43*(2), 618–644. https://doi.org/10.1016/j.dss.2005.05.019

Kässi, O., & Lehdonvirta, V. (2018). Online labour index: Measuring the online gig economy for policy and research. *Technological Forecasting and Social Change, 137,* 241–248. https://doi.org/10.1016/j.techfore.2018.07.056

Kenney, M., & Zysman, J. (2016). The rise of the platform economy. *Issues in Science and Technology, 32*(3), 61–69.

Kittur, A., Nickerson, J. V., Bernstein, M., Gerber, E., Shaw, A., Zimmerman, J., Lease, M., & Horton, J. (2013). The future of crowd work. In *Proceedings of the 2013 Conference on Computer-supported Cooperative Work* (CSCW '13). New York, NY, USA: ACM Press. https://doi.org/10.1145/2441776.2441923

Kossek, E. E. (2016). Managing work-life boundaries in the digital age. *Organizational Dynamics, 45*(3), 258–270. https://doi.org/10.1016/j.orgdyn.2016.07.010

Kossek, E. E., Lautsch, B. A., & Eaton, S. C. (2006). Telecommuting, control, and boundary management: Correlates of policy use and practice, job control, and work-family effectiveness. *Journal of Vocational Behavior, 68*(2), 347–367. https://doi.org/10.1016/j.jvb.2005.07.002

Kuek, S. C., Paradi-Guilford, C., Fayomi, T., Imaizumi, S., Ipeirotis, P., Pina, P., & Singh, M. (2015), *The global opportunity in online outsourcing.*

Langley, P., & Leyshon, A. (2017). Platform capitalism: The intermediation and capitalization of digital economic circulation. *Finance and Society, 3*(1), 11–31. https://doi.org/10.2218/finsoc.v3i1.1936

Lehdonvirta, V. (2018). Flexibility in the gig economy: Managing time on three online piecework platforms. *New Technology, Work and Employment, 33*(1), 13–29. https://doi.org/10.1111/ntwe.12102

Leimeister, J. M., Zogaj, S., Durward, D., & Blohm, I. (2016). *Systematisierung und Analyse von Crowd-Sourcing-Anbietern und Crowd-Work-Projekten, Study der Hans-Böckler-Stiftung* (Vol. 324). Düsseldorf: Hans-Böckler-Stiftung.

Meyer, S.-C., Tisch, A., & Hünefeld, L. (2019). Arbeitsintensivierung und Handlungsspielraum in digitalisierten Arbeitswelten – Herausforderung für das Wohlbefinden von Beschäftigten? Industrielle Beziehungen. *Zeitschrift Für Arbeit, Organisation Und Management, 26*(2-2019), 207–231. https://doi.org/10.3224/indbez.v26i2.06

Mrass, V., Peters, C., & Leimeister, J. M. (2020). Crowdworking platforms in Germany: Business insights from a study & implications for society. *SSRN Electronic Journal.* https://doi.org/10.2139/ssrn.3926265

Peer, E., Vosgerau, J., & Acquisti, A. (2014). Reputation as a sufficient condition for data quality on Amazon Mechnical Turk. *Behavior Research Methods, 46*(4), 1023–1031. https://doi.org/10.3758/s13428-013-0434-y

Pesole, A., Urzí Brancati, M. C., Fernández-Macías, E., Biagi, F., & González Vázquez, I. (2018). *Platform workers in Europe.* Luxembourg: Publications Office of the European Union.

Pongratz, H. J., & Bormann, S. (2017). Online-arbeit auf internet-plattformen: Empirische befunde zum 'crowdworking' in Deutschland. Arbeits- und Industriesoziologische Studien, *10*(2), 158–181. https://doi.org/10.21241/ssoar.64850

Quinlan, M., Mayhew, C., & Bohle, P. (2001). The global expansion of precarious employment, work disorganization, and consequences for occupational health: A review of recent research. *International Journal of Health Services: Planning, Administration, Evaluation, 31*(2), 335–414. https://doi.org/10.2190/607H-TTV0-QCN6-YLT4

Reimann, M., & Abendroth, A. (2023). Flexible working and its relation with work-life conflicts and well-being among crowdworkers in Germany. *WORK: A Journal of Prevention, Assessment & Rehabilitation, 74*(2).

Riedl, C., & Seidel, V. P. (2018). Learning from mixed signals in online innovation communities. *Organization Science, 29*(6), 1010–1032. https://doi.org/10.1287/orsc.2018.1219

Schieman, S., & Glavin, P. (2008). Trouble at the border?: Gender, flexibility at work, and the work-home interface. *Social Problems, 55*(4), 590–611. https://doi.org/10.1525/sp.2008.55.4.590

Schlicher, K. D., Schulte, J., Reimann, M., & Maier, G. W. (2021). Flexible, self-determined… and unhealthy? An empirical study on somatic health among crowdworkers. *Frontiers in Psychology, 12.* https://doi.org/10.3389/fpsyg.2021.724966

Schneider, M. R., & Hemsen, P. (2021). Freelancers who stay? *A fuzzy-set qualitative comparative analysis of affective and calculative commitment among crowdworkers to a platform.*

Schörpf, P., Flecker, J., Schönauer, A., & Eichmann, H. (2017). Triangular love-hate: Management and control in creative crowdworking. *New Technology, Work and Employment, 32*(1), 43–58. https://doi.org/10.1111/ntwe.12080

Schulte, J., Schlicher, K. D., & Maier, G. W. (2020). Working everywhere and every time? Chances and risks in crowdworking and crowdsourcing work design. *Gruppe Interaktion Organisation Zeitschrift für Angewandte Organisationspsychologie (GIO), 51*(1), 59–69. https://doi.org/10.1007/s11612-020-00503-3

Schulten, M. B., & Schaefer, F. (2015). Affective commitment and customer loyalty in crowdsourcing: Antecedents, interdependencies, and practical implications. *The International Review of Retail, Distribution and Consumer Research, 25*(5), 516–528. https://doi.org/10.1080/09593969.2015.1081099

Serfling, O. (2019). Crowdworking monitor No. 2. *Hochschule Rhein-Waal.* https://doi.org/10.13140/RG.2.2.36135.91044

Sverke, M., Hellgren, J., & Näswall, K. (2002). No security: A meta-analysis and review of job insecurity and its consequences. *Journal of Occupational Health Psychology, 7*(3), 242–264. https://doi.org/10.1037/1076-8998.7.3.242

Tavares, A. I. (2017). Telework and health effects review. *International Journal of Healthcare, 3*(2), 30. https://doi.org/10.5430/ijh.v3n2p30

Taylor, J., & Joshi, K. D. (2019). Joining the crowd: The career anchors of information technology workers participating in crowdsourcing. *Information Systems Journal, 29*(3), 641–673. https://doi.org/10.1111/isj.12225

Warren, T. (2021). Work-life balance and gig work: 'Where are we now' and 'where to next' with the work-life balance agenda? *Journal of Industrial Relations, 63*(4), 522–545. https://doi.org/10.1177/00221856211007161

Wood, A. J., Graham, M., Lehdonvirta, V., & Hjorth, I. (2019). Good gig, bad gig: Autonomy and algorithmic control in the global gig economy. *Work, Employment and Society, 33*(1), 56–75. https://doi.org/10.1177/0950017018785616

Zuchowski, O., Posegga, O., Schlagwein, D., & Fischbach, K. (2016). Internal crowdsourcing: Conceptual framework, structured review, and research agenda. *Journal of Information Technology, 31*(2), 166–184. https://doi.org/10.1057/jit.2016.14

Zyskowski, K., Morris, M. R., Bigham, J. P., Gray, M. L., & Kane, S. K (2015). Accessible crowdwork? In D. Cosley, A. Forte, L. Ciolfi, D. McDonald (Eds.), *Proceedings of the 18th ACM Conference on Computer Supported Cooperative Work & Social Computing* (pp. 1682–1693). New York, NY, USA: ACM. https://doi.org/10.1145/2675133.2675158

Printed in the United States
by Baker & Taylor Publisher Services